Marx Joyce
Hardy Austen
Abbott Machiavelli Chesterton Hugo
Defoe Melville Montaigne Cooper Emerson Eliot Grimm
Stoker Carroll Haggard Molière
Wilde Christie Byron Schiller
Maupassant
Garnett Fitzgerald Engels
Goethe Einstein Hawthorne Kafka
Cotton Dostoyevsky Smith
Baum Kipling Doyle Hall
Dumas Henry Willis
Leslie Flaubert Nietzsche
Stockton Turgenev Balzac
Burroughs Vatsyayana Crane
Curtis Tocqueville Verne
Homer Widger Gogol Vinci
Darwin Tolstoy Busch
Potter Thoreau Whitman
Freud Zola Twain Scott
Kant Jowett Lawrence Plato Harte
Stevenson Dickens Hesse
Andersen Burton
London Descartes Voltaire
Poe Aristotle Wells Cervantes
Hale James Hastings Cooke
Bunner Shakespeare
Richter Chambers Irving
Doré da Benedict
Swift Dante Shaw Pushkin Alcott
Chekhov Wodehouse
Newton

tredition®

tredition was established in 2006 by Sandra Latusseck and Soenke Schulz. Based in Hamburg, Germany, tredition offers publishing solutions to authors and publishing houses, combined with worldwide distribution of printed and digital book content. tredition is uniquely positioned to enable authors and publishing houses to create books on their own terms and without conventional manufacturing risks.

For more information please visit: www.tredition.com

TREDITION CLASSICS

This book is part of the TREDITION CLASSICS series. The creators of this series are united by passion for literature and driven by the intention of making all public domain books available in printed format again - worldwide. Most TREDITION CLASSICS titles have been out of print and off the bookstore shelves for decades. At tredition we believe that a great book never goes out of style and that its value is eternal. Several mostly non-profit literature projects provide content to tredition. To support their good work, tredition donates a portion of the proceeds from each sold copy. As a reader of a TREDITION CLASSICS book, you support our mission to save many of the amazing works of world literature from oblivion. See all available books at www.tredition.com.

 Project Gutenberg

The content for this book has been graciously provided by Project Gutenberg. Project Gutenberg is a non-profit organization founded by Michael Hart in 1971 at the University of Illinois. The mission of Project Gutenberg is simple: To encourage the creation and distribution of eBooks. Project Gutenberg is the first and largest collection of public domain eBooks.

Flowers and Flower-Gardens With an Appendix of Practical Instructions and Useful

David Lester Richardson

Imprint

This book is part of TREDITION CLASSICS

Author: David Lester Richardson
Cover design: Buchgut, Berlin – Germany

Publisher: tredition GmbH, Hamburg - Germany
ISBN: 978-3-8424-4963-3

www.tredition.com
www.tredition.de

Copyright:
The content of this book is sourced from the public domain.

The intention of the TREDITION CLASSICS series is to make world literature in the public domain available in printed format. Literary enthusiasts and organizations, such as Project Gutenberg, worldwide have scanned and digitally edited the original texts. tredition has subsequently formatted and redesigned the content into a modern reading layout. Therefore, we cannot guarantee the exact reproduction of the original format of a particular historic edition. Please also note that no modifications have been made to the spelling, therefore it may differ from the orthography used today.

FLOWERS AND FLOWER-GARDENS.

BY

DAVID LESTER RICHARDSON,

PRINCIPAL OF THE HINDU METROPOLITAN COLLEGE, AND AUTHOR OF "LITERARY LEAVES," "LITERARY RECREATIONS," &C.

WITH AN APPENDIX OF

PRACTICAL INSTRUCTIONS AND USEFUL INFORMATION RESPECTING THE ANGLO- INDIAN FLOWER-GARDEN.

CALCUTTA:

MDCCCLV.

PREFACE.

> In every work regard the writer's end, Since none can compass more than they intend.

Pope.

This volume is far indeed from being a scientific treatise *On Flowers and Flower-Gardens*:--it is mere gossip in print upon a pleasant subject. But I hope it will not be altogether useless. If I succeed in my object I shall consider that I have gossipped to some purpose. On several points--such as that of the mythology and language of flowers--I have said a good deal more than I should have done had I been writing for a different community. I beg the London critics to bear this in mind. I wished to make the subject as attractive as possible to some classes of people here who might not have been disposed to pay any attention to it whatever if I had not studied their amusement as much as their instruction. I have tried to sweeten the edge of the cup.

I did not at first intend the book to exceed fifty pages: but I was almost insensibly carried on further and further from the proposed limit by the attractive nature of the materials that pressed upon my notice. As by far the largest portion, of it has been written hurriedly, amidst other avocations, and bit by bit; just as the Press demanded an additional supply of "*copy*," I have but too much reason to apprehend that it will seem to many of my readers, fragmentary and ill-connected. Then again, in a city like Calcutta, it is not easy to prepare any thing satisfactorily that demands much literary or scientific research. There are very many volumes in all the London Catalogues, but not immediately obtainable in Calcutta, that I should have been most eager to refer to for interesting and valuable information, if they had been at hand. The mere titles of these books have often tantalized me with visions of riches beyond my reach. I might indeed have sent for some of these from England, but I had announced this volume, and commenced the printing of it, before it occurred to me that it would be advisable to extend the matter beyond the limits I had originally contemplated. I must now send it forth, "with all its imperfections on its head;" but not without the hope that in spite of these, it will be found calculated to increase the taste amongst my brother exiles here for flowers and flower-gardens, and lead many of my Native friends--(particularly those who have been educated at the Government Colleges,--who have imbibed some English thoughts and feelings--and who are so fortunate as to be in possession of landed property)--to improve their parterres,--and set an example to their poorer countrymen of that neatness and care and cleanliness and order which may make even the peasant's cottage and the smallest plot of ground assume an aspect of comfort, and afford a favorable indication of the character of the possessor.

D.L.R.

Calcutta, September 21st 1855.

ERRATA.

A friend tells me that the allusion to the Acanthus on the first page of this book is obscurely expressed, that it was not the *root* but the *leaves* of the plant that suggested the idea of the Corinthian capital. The root of the Acanthus produced the leaves which overhanging the sides of the basket struck the fancy of the Architect. This was, indeed, what I *meant* to say, and though I have not very lucidly expressed myself, I still think that some readers might have understood me rightly even without the aid of this explanation, which, however, it is as well for me to give, as I wish to be intelligible to *all*. A writer should endeavor to make it impossible for any one to misapprehend his meaning, though there are some writers of high name both in England and America who seem to delight in puzzling their readers.

At the bottom of page 200, allusion is made to the dotted lines at some of the open turns in the engraved labyrinth. By some accident or mistake the dots have been omitted, but any one can understand where the stop hedges which the dotted lines indicated might be placed so as to give the wanderer in the maze, additional trouble to find his way out of it.

ON FLOWERS AND FLOWER-GARDENS,

> For lo, the winter is past, the rain is over and gone; the flowers appear on the earth, the time of the singing of birds is come, and the voice of the turtle is heard in our land.

The Song of Solomon.

> These are thy glorious works, Parent of good! Almighty, Thine this universal frame, Thus wondrous fair; Thyself how wondrous then!

Milton.

> Soft roll your incense, herbs and fruits and flowers, In mingled clouds to HIM whose sun exalts Whose breath perfumes you, and whose pencil paints.

Thomson.

A taste for floriculture is spreading amongst Anglo-Indians. It is a good sign. It would be gratifying to learn that the same refining taste had reached the Natives also--even the lower classes of them. It is a cheap enjoyment. A mere palm of ground may be glorified by a few radiant blossoms. A single clay jar of the rudest form may be so enriched and beautified with leaves and blossoms as to fascinate the eye of taste. An old basket, with a broken tile at the top of it, and the root of the acanthus within, produced an effect which seemed to Calimachus, the architect, "the work of the Graces." It suggested the idea of the capital of the Corinthian column, the most elegant architectural ornament that Art has yet conceived.

Flowers are the poor man's luxury; a refinement for the uneducated. It has been prettily said that the melody of birds is the poor man's music, and that flowers are the poor man's poetry. They are "a discipline of humanity," and may sometimes ameliorate even a coarse and vulgar nature, just as the cherub faces of innocent and happy children are sometimes found to soften and purify the corrupted heart. It would be a delightful thing to see the swarthy cottagers of India throwing a cheerful grace on their humble sheds and small plots of ground with those natural embellishments which no productions of human skill can rival.

The peasant who is fond of flowers--if he begin with but a dozen little pots of geraniums and double daisies upon his window sills, or with a honeysuckle over his humble porch--gradually acquires a habit, not only of decorating the outside of his dwelling and of cultivating with care his small plot of ground, but of setting his house in order within, and making every thing around him agreeable to the eye. A love of cleanliness and neatness and simple ornament is a moral feeling. The country laborer, or the industrious mechanic, who has a little garden to be proud of, the work of his own hand, becomes attached to his place of residence, and is perhaps not only a better subject on that account, but a better neighbour--a better man. A taste for flowers is, at all events, infinitely preferable to a taste for the excitements of the pot-house or the tavern or the turf or the gaming table, or even the festal board, especially for people of feeble health--and above all, for the poor--who should endeavor to satisfy themselves with inexpensive pleasures.[001]

In all countries, civilized or savage, and on all occasions, whether of grief or rejoicing, a natural fondness for flowers has been exhibited, with more or less tenderness or enthusiasm. They beautify religious rites. They are national emblems: they find a place in the blazonry of heraldic devices. They are the gifts and the language of friendship and of love.

Flowers gleam in original hues from graceful vases in almost every domicile where Taste presides; and the hand of "nice Art" charms us with "counterfeit presentments" of their forms and colors, not only on the living canvas, but even on our domestic Chinaware, and our mahogany furniture, and our wall-papers and hangings and carpets, and on our richest apparel for holiday occasions and our simplest garments for daily wear. Even human Beauty, the Queen of all loveliness on earth, engages Flora as her handmaid at the toilet, in spite of the dictum of the poet of 'The Seasons,' that "Beauty when unadorned is adorned the most."

Flowers are hung in graceful festoons both in churches and in ball- rooms. They decorate the altar, the bride-bed, the cradle, and the bier. They grace festivals, and triumphs, and processions; and cast a glory on gala days; and are amongst the last sad honors we pay to the objects of our love.

I remember the death of a sweet little English girl of but a year old, over whom, in her small coffin, a young and lovely mother sprinkled the freshest and fairest flowers. The task seemed to soften--perhaps to sweeten--her maternal grief. I shall never forget the sight. The bright-hued blossoms seemed to make her oblivious for a moment of the darkness and corruption to which she was so soon to consign her priceless treasure. The child's sweet face, even in death, reminded me that the flowers of the field and garden, however lovely, are all outshone by human beauty. What floral glory of the wild-wood, or what queen of the parterre, in all the pride of bloom, laughing in the sun-light or dancing in the breeze, hath a charm that could vie for a single moment with the soft and holy lustre of that motionless and faded human lily? I never more deeply felt the force of Milton's noble phrase "*the human face divine*" than when gazing on that sleeping child. The fixed placid smile, the smoothly closed eye with its transparent lid, the air of profound tranquillity, the simple purity (elevated into an aspect of bright intelligence, as if the little cherub already experienced the beatitude of another and a better world,) were perfectly angelic--and mocked all attempt at description. "Of such is the Kingdom of Heaven!"

O flower of an earthly spring! destined to blossom in the eternal summer of another and more genial region! Loveliest of lovely children-- loveliest to the last! More beautiful in death than aught still living! Thou seemest now to all who miss and mourn thee but a sweet name--a fair vision--a precious memory;--but in reality thou art a more truly living thing than thou wert before or than aught thou hast left behind. Thou hast come early into a rich inheritance. Thou hast now a substantial existence, a genuine glory, an everlasting possession, beyond the sky. Thou hast exchanged the frail flowers that decked thy bier for amaranthine hues and fragrance, and the brief and uncertain delights of mortal being for the eternal and perfect felicity of angels!

I never behold elsewhere any of the specimens of the several varieties of flowers which the afflicted parent consigned to the hallowed little coffin without recalling to memory the sainted child taking her last rest on earth. The mother was a woman of taste and sensibility, of high mind and gentle heart, with the liveliest sense of the loveliness of all lovely things; and it is hardly necessary to remind the reader how much refinement such as hers may sometimes alleviate the severity of sorrow.

Byron tells us that the stars are

> A beauty and a mystery, and create In us such love and reverence from afar That fortune, fame, power, life, have named themselves *a star*.

But might we not with equal justice say that every thing excellent and beautiful and precious has named itself *a flower*?

If stars teach as well as shine--so do flowers. In "still small accents" they charm "the nice and delicate ear of thought" and sweetly whisper that "the hand that made them is divine."

The stars are the poetry of heaven--the clouds are the poetry of the middle sky--the flowers are the poetry of the earth. The last is the loveliest to the eye and the nearest to the heart. It is incomparably the sweetest external poetry that Nature provides for man. Its attractions are the most popular; its language is the most intelligible. It is of all others the best adapted to every variety and degree of mind. It is the most endearing, the most familiar, the most homefelt, and congenial. The stars are for the meditation of poets and philosophers; but flowers are not exclusively for the gifted or the scientific; they are the property of all. They address themselves to our common nature. They are equally the delight of the innocent little prattler and the thoughtful sage. Even the rude unlettered rustic betrays some feeling for the beautiful in the presence of the lovely little community of the field and garden. He has no sympathy for the stars: they are too mystical and remote. But the flowers as they blush and smile beneath his eye may stir the often deeply hidden lovingness and gentleness of his nature. They have a social and domestic aspect to which no one with a human heart can be quite indifferent. Few can doat upon the distant flowers of the sky as many of us doat upon the flowers at our feet. The stars are wholly independent of man: not so the sweet children of Flora. We tend upon and cherish them with a parental pride. They seem especially meant for man and man for them. They often need his kindest nursing. We place them with guardian hand in the brightest light and the most wholesome air. We quench with liquid life their sun-raised thirst, or shelter them from the wintry blast, or prepare and enrich their nutritious beds. As they pine or prosper they agitate us with tender anxieties, or thrill us with exultation and delight. In the little plot of ground that fronts an English cottage the flowers are like members of the household. They are of the same family. They are almost as lovely as the children that play with them--though their happy human associates may be amongst

The sweetest things that ever grew Beside a human door.

The Greeks called flowers the *Festival of the eye*: and so they are: but they are something else, and something better.

A flower is not a flower alone, A thousand sanctities invest it.

Flowers not only touch the heart; they also elevate the soul. They bind us not entirely to earth; though they make earth delightful. They attract our thoughts downward to the richly embroidered ground only to raise them up again to heaven. If the stars are the scriptures of the sky, the flowers are the scriptures of the earth. If the stars are a more glorious revelation of the Creator's majesty and might, the flowers are at least as sweet a revelation of his gentler attributes. It has been observed that

An undevout astronomer is mad.

The same thing may be said of an irreverent floriculturist, and with equal truth--perhaps indeed with greater. For the astronomer, in some cases, may be hard and cold, from indulging in habits of thought too exclusively mathematical. But the true lover of flowers has always something gentle and genial in his nature. He never looks upon his floral-family without a sweetened smile upon his face and a softened feeling in his heart; unless his temperament be strangely changed and his mind disordered. The poets, who, speaking generally, are constitutionally religious, are always delighted readers of the flower- illumined pages of the book of nature. One of these disciples of Flora earnestly exclaims:

> Were I, O God, in churchless lands remaining Far from all voice of teachers and divines, My soul would find in flowers of thy ordaining Priests, sermons, shrines

The popular little preachers of the field and garden, with their lovely faces, and angelic language--sending the while such ambrosial incense up to heaven--insinuate the sweetest truths into the human heart. They lead us to the delightful conclusion that beauty is in the list of the *utilities*--that the Divine Artist himself is *a lover of loveliness*-- that he has communicated a taste for it to his creatures and most lavishly provided for its gratification.

> Not a flower But shows some touch, in freckle, streak or stain, Of His unrivalled pencil. He inspires Their balmy odours, and imparts then hues, And bathes their eyes with nectar, and includes In grains as countless as the sea side sands The forms with which he sprinkles all the earth.

Cowper.

In the eye of Utilitarianism the flowers are but idle shows. God might indeed have made this world as plain as a Quaker's garment, without retrenching one actual necessary of physical existence; but He has chosen otherwise; and no earthly potentate was ever so richly clad as his mother earth. "Behold the lilies of the field, they spin not, neither do they toil, yet Solomon in all his glory was not arrayed like one of these!" We are thus instructed that man was not meant to live by bread alone, and that the gratification of a sense of beauty is equally innocent and natural and refining. The rose is permitted to spread its sweet leaves to the air and dedicate its beauty to the sun, in a way that is quite perplexing to bigots and stoics and political economists. Yet God has made nothing in vain! The Great Artist of the Universe must have scattered his living hues and his forms of grace over the surface of the earth for some especial and worthy purpose. When Voltaire was congratulated on the rapid growth of his plants, he observed that "*they had nothing else to do.*" Oh, yes--they had something else to do,--they had to adorn the earth, and to charm the human eye, and through the eye to soften and cheer the heart and elevate the soul!

I have often wished that Lecturers on Botany, instead of confining their instructions to the mere physiology, or anatomy, or classification or nomenclature of their favorite science, would go more into the poetry of it, and teach young people to appreciate the moral influences of the floral tribes--to draw honey for the human heart from the sweet breasts of flowers--to sip from their radiant chalices a delicious medicine for the soul.

Flowers are frequently hallowed by associations far sweeter than their sweetest perfume. "I am no botanist:" says Southey in a letter to Walter Savage Landor, "but like you, my earliest and best recollections are connected with flowers, and they always carry me back to other days. Perhaps this is because they are the only things which affect our senses precisely as they did in our childhood. The sweetness of the violet is always the same; and when you rifle a rose and drink, as it were, its fragrance, the refreshment is the same to the old man as to the boy. Sounds recal the past in the same manner, but they do not bring with them individual scenes like the cowslip field, or the corner of the garden to which we have transplanted field-flowers."

George Wither has well said in commendation of his Muse:

> Her divine skill taught me this; That from every thing I saw I could some instruction draw, And raise pleasure to the height By the meanest object's sight, By the murmur of a spring *Or the least bough's rustelling; By a daisy whose leaves spread Shut, when Titan goes to bed; Or a shady bush or tree*, She could more infuse in me Than all Nature's beauties can In some other wiser man.

We must not interpret the epithet *wiser* too literally. Perhaps the poet speaks ironically, or means by some other *wiser man*, one allied in character and temperament to a modern utilitarian Philosopher.

Wordsworth seems to have had the lines of George Wither in his mind when he said

> Thanks to the human heart by which we live, Thanks to its tenderness, its joys, and fears, To me the meanest flower that blows can give Thoughts that do often lie too deep for tears.

Thomas Campbell, with a poet's natural gallantry, has exclaimed,

> Without the smile from partial Beauty won, Oh! what were man?--a world without a sun!

Let a similar compliment be presented to the "painted populace that dwell in fields and lead ambrosial lives." What a desert were this scene without its flowers--it would be like the sky of night without its stars! "The disenchanted earth" would "lose her lustre." Stars of the day! Beautifiers of the world! Ministrants of delight! Inspirers of kindly emotions and the holiest meditations! Sweet teachers of the serenest wisdom! So beautiful and bright, and graceful, and fragrant--it is no marvel that ye are equally the favorites of the rich and the poor, of the young and the old, of the playful and the pensive!

Our country, though originally but sparingly endowed with the living jewelry of nature, is now rich in the choicest flowers of all other countries.

> Foreigners of many lands, They form one social shade, as if convened By magic summons of the Orphean lyre.

Cowper.

These little "foreigners of many lands" have been so skilfully acclimatized and multiplied and rendered common, that for a few shillings an English peasant may have a parterre more magnificent than any ever gazed upon by the Median Queen in the hanging gardens of Babylon. There is no reason, indeed, to suppose that even the first parents of mankind looked on finer flowers in Paradise itself than are to be found in the cottage gardens that are so thickly distributed over the hills and plains and vallies of our native land.

> The red rose, is the red rose still, and from the lily's cup An odor fragrant as at first, like frankincense goes up.

Mary Howitt.

Our neat little gardens and white cottages give to dear old England that lovely and cheerful aspect, which is so striking and attractive to her foreign visitors. These beautiful signs of a happy political security and individual independence and domestic peace and a love of order and a homely refinement, are scattered all over the land, from sea to sea. When Miss Sedgwick, the American authoress, visited England, nothing so much surprised and delighted her as the gay flower-filled gardens of our cottagers. Many other travellers, from almost all parts of the world, have experienced and expressed the same sensations on visiting our shores, and it would be easy to compile a voluminous collection of their published tributes of admiration. To a foreign visitor the whole country seems a garden--in the words of Shakespeare--"a *sea-walled garden.*"

In the year 1843, on a temporary return to England after a long Indian exile, I travelled by railway for the first time in my life. As I glided on, as smoothly as in a sledge, over the level iron road, with such magical rapidity--from the pretty and cheerful town of Southampton to the greatest city of the civilized world--every thing was new to me, and I gave way to child-like wonder and child-like exultation.[002] What a quick succession of lovely landscapes greeted the eye on either side? What a garden-like air of universal cultivation! What beautiful smooth slopes! What green, quiet meadows! What rich round trees, brooding over their silent shadows! What exquisite dark nooks and romantic lanes! What an aspect of unpretending happiness in the clean cottages, with their little trim gardens! What tranquil grandeur and rural luxury in the noble mansions and glorious parks of the British aristocracy! How the love of nature thrilled my heart with a gentle and delicious agitation, and how proud I felt of my dear native land! It is, indeed, a fine thing to be an Englishman. Whether at home or abroad, he is made conscious of the claims of his country to respect and admiration. As I fed my eyes on the loveliness of Nature, or turned to the miracles of Art and Science on every hand, I had always in my mind a secret reference to the effect which a visit to England must produce upon an intelligent and observant foreigner.

> Heavens! what a goodly prospect spreads around Of hills and dales and woods and lawns and spires, And glittering towns and gilded streams, 'till all The stretching landscape into smoke decays! Happy Brittannia! where the Queen of Arts, Inspiring vigor, Liberty, abroad Walks unconfined, even to thy farthest cots, And scatters plenty with unsparing hand.

Thomson.

And here let me put in a word in favor of the much-abused English climate. I cannot echo the unpatriotic discontent of Byron when he speaks of

> The cold and cloudy clime Where he was born, but where he would not die.

Rather let me say with the author of "*The Seasons,*" in his address to England.

> Rich is thy soil and merciful thy clime.

King Charles the Second when he heard some foreigners condemning our climate and exulting in their own, observed that in his opinion that was the best climate in which a man could be out in the open air with pleasure, or at least without trouble and inconvenience, the most days of the year and the most hours of the day; and this he held was the case with the climate of England more than that of any other country in Europe. To say nothing of the lovely and noble specimens of human nature to which it seems so congenial, I may safely assert that it is peculiarly favorable, with, rare exceptions, to the sweet children of Flora. There is no country in the world in which there are at this day such innumerable tribes of flowers. There are in England two thousand varieties of the rose alone, and I venture to express a doubt whether the richest gardens of Persia or Cashmere could produce finer specimens of that universal favorite than are to be found in some of the small but highly cultivated enclosures of respectable English rustics.

The actual beauty of some of the commonest flowers in our gardens can be in no degree exaggerated--even in the daydreams of the most inspired poet. And when the author of Lalla Rookh talks so musically and pleasantly of the fragrant bowers of Amberabad, the country of Delight, a Province in Jinnistan or Fairy Land, he is only thinking of the shrubberies and flower-beds at Sloperton Cottage, and the green hills and vales of Wiltshire.

Sir William Temple observes that "besides the temper of our climate there are two things particular to us, that contribute much to the beauty and elegance of our gardens--which are, *the gravel of our walks and the fineness and almost perpetual greenness of our turf*."

"The face of England is so beautiful," says Horace Walpole, "that I do not believe that Tempe or Arcadia was half so rural; for both lying in hot climates must have wanted *the moss of our gardens*." Meyer, a German, a scientific practical gardener, who was also a writer on gardening, and had studied his art in the Royal Gardens at Paris, and afterwards visited England, was a great admirer of English Gardens, but despaired of introducing our style of gardening into Germany, *chiefly on account of its inferior turf for lawns*. "Lawns and gravel walks," says a writer in the *Quarterly Review*, "are the pride of English Gardens," "The smoothness and verdure of our lawns," continues the same writer, "is the first thing in our gardens that catches the eye of a foreigner; the next is the fineness and firmness of our gravel walks." Mr. Charles Mackintosh makes the same observation. "In no other country in the world," he says, "do such things exist." Mrs. Stowe, whose *Uncle Tom* has done such service to the cause of liberty in America, on her visit to England seems to have been quite as much enchanted with our scenery, as was her countrywoman, Miss Sedgwick. I am pleased to find Mrs. Stowe recognize the superiority of English landscape-gardening and of our English verdure. She speaks of, "the princely art of landscape- gardening, for which England is so famous," and of "*vistas of verdure and wide sweeps of grass, short, thick, and vividly green* as the velvet moss sometimes seen growing on rocks in new England." "Grass," she observes, "is an art and a science in England--it is an institution. The pains that are taken in sowing, tending, cutting, clipping, rolling and otherwise nursing and coaxing it, being seconded by the often-falling tears of the climate, produce results which must be seen to be appreciated." This is literally true: any sight more inexpressibly exquisite than that of an English lawn in fine order is what I am quite unable to conceive.[003]

I recollect that in one of my visits to England, (in 1827) I attempted to describe the scenery of India to William Hazlitt--not the living son but the dead father. Would that he were still in the land of the living by the side of his friend Leigh Hunt, who has been pensioned by the Government for his support of that cause for which they were both so bitterly persecuted by the ruling powers in days gone by. I flattered myself into the belief that Hazlitt was interested in some of my descriptions of Oriental scenes. What moved him most was an account of the dry, dusty, burning, grassless plains of Bundelcund in the hot season. I told him how once while gasping for breath in a hot verandah and leaning over the rails I looked down upon the sun-baked ground.

> "A change came o'er the spirit of my dream."

I suddenly beheld with all the distinctness of reality the rich, cool, green, unrivalled meads of England. But the vision soon melted away, and I was again in exile. I wept like a child. It was like a beautiful mirage of the desert, or one of those waking dreams of home which have sometimes driven the long-voyaging seaman to distraction and urged him by an irresistible impulse to plunge headlong into the ocean.

When I had once more crossed the wide Atlantic--and (not by the necromancy of imagination but by a longer and more tedious transit) found myself in an English meadow,--I exclaimed with the poet,

> Thou art free My country! and 'tis joy enough and pride For one hour's perfect bliss, *to tread the grass Of England once again*.

I felt my childhood for a time renewed, and was by no means disposed to second the assertion that

> "Nothing can bring back the hour Of splendour in the grass, of glory in the flower."

I have never beheld any thing more lovely than scenery characteristically English; and Goldsmith, who was something of a traveller, and had gazed on several beautiful countries, was justified in speaking with such affectionate admiration of our still more beautiful England,

> Where lawns extend that scorn Arcadian pride.

It is impossible to put into any form of words the faintest representation of that delightful summer feeling which, is excited in fine weather by the sight of the mossy turf of our country. It is sweet indeed to go,

> Musing through the *lawny* vale:

alluded to by Warton, or over Milton's "level downs," or to climb up Thomson's

> Stupendous rocks That from the sun-redoubling valley lift Cool to the middle air their *lawny* tops.

It gives the Anglo-Indian Exile the heart-ache to think of these ramblings over English scenes.

ENGLAND.

> Bengala's plains are richly green, Her azure skies of dazzling sheen, Her rivers vast, her forests grand. Her bowers brilliant,--but the land, Though dear to countless eyes it be, And fair to mine, hath not for me The charm ineffable of *home*; For still I yearn to see the foam Of wild waves on thy pebbled shore, Dear Albion! to ascend once more Thy snow-white cliffs; to hear again The murmur of thy circling main-- To

stroll down each romantic dale Beloved in boyhood--to inhale Fresh life on green and breezy hills-- To trace the coy retreating rills-- To see the clouds at summer-tide Dappling all the landscape wide-- To mark the varying gloom and glow As the seasons come and go-- Again the green meads to behold Thick strewn with silvery gems and gold, Where kine, bright-spotted, large, and sleek, Browse silently, with aspect meek, Or motionless, in shallow stream Stand mirror'd, till their twin shapes seem, Feet linked to feet, forbid to sever, By some strange magic fixed for ever. And oh! once more I fain would see (Here never seen) a poor man *free*,[004] And valuing more an humble name, But stainless, than a guilty fame, How sacred is the simplest cot, Where Freedom dwells!-- where she is not How mean the palace! Where's the spot She loveth more than thy small isle, Queen of the sea? Where hath her smile So stirred man's inmost nature? Where Are courage firm, and virtue fair, And manly pride, so often found As in rude huts on English ground, Where e'en the serf who slaves for hire May kindle with a freeman's fire? How proud a sight to English eyes Are England's village families! The patriarch, with his silver hair, The matron grave, the maiden fair. The rose-cheeked boy, the sturdy lad, On Sabbath day all neatly clad:-- Methinks I see them wend their way On some refulgent morn of May, By hedgerows trim, of fragrance rare, Towards the hallowed House of Prayer! I can love *all* lovely lands, But England *most*; for she commands. As if she bore a parent's part, The dearest movements of my heart; And here I may not breathe her name. Without a thrill through all my frame. Never shall this heart be cold To thee, my country! till the mould (Or *thine* or *this*) be o'er it spread. And form its dark and silent bed. I never think of bliss below But thy sweet hills their green heads show, Of love and beauty never dream. But English faces round me gleam!

D.L.R.

I have often observed that children never wear a more charming aspect than when playing in fields and gardens. In another volume I have recorded some of my impressions respecting the prominent interest excited by these little flowers of humanity in an English landscape.

THE RETURN TO ENGLAND.

When I re-visited my dear native country, after an absence of many weary years, and a long dull voyage, my heart was filled with unutterable delight and admiration. The land seemed a perfect paradise. It was in the spring of the year. The blue vault of heaven--the clear atmosphere-- the balmy vernal breeze--the quiet and picturesque cattle, browsing on luxuriant verdure, or standing knee deep in a crystal lake--the hills sprinkled with snow-white sheep and sometimes partially shadowed by a wandering cloud--the meadows glowing with golden butter-cups and be- dropped with daisies--the trim hedges of crisp and sparkling holly--the sound of near but unseen rivulets, and the songs of foliage-hidden birds--the white cottages almost buried amidst trees, like happy human nests--the ivy-covered church, with its old grey spire "pointing up to heaven," and its gilded vane gleaming in the light--the sturdy peasants with their instruments of healthy toil--the white-capped matrons bleaching their newly-washed garments in the sun, and throwing them like snow-patches on green slopes, or glossy garden shrubs-- the sun-browned village girls, resting idly on their round elbows at small open casements, their faces in sweet keeping with the trellised flowers:--all formed a combination of enchantments that would mock the happiest imitative efforts of human art. But though the bare enumeration of the details of this English picture, will, perhaps, awaken many dear recollections in the reader's mind, I have omitted by far the most interesting feature of the whole scene-- *the rosy children, loitering about the cottage gates, or tumbling gaily on the warm grass.*[005][006]

Two scraps of verse of a similar tendency shall follow this prose description:--

AN ENGLISH LANDSCAPE.

> I stood, upon an English hill, And saw the far meandering rill, A vein of liquid silver, run Sparkling in the summer sun; While adown that green hill's side, And along the valley wide, Sheep, like small clouds touched with light, Or like little

breakers bright, Sprinkled o'er a smiling sea, Seemed to float at liberty. Scattered all around were seen, White cots on the meadows green. Open to the sky and breeze, Or peeping through the sheltering trees, On a light gate, loosely hung, Laughing children gaily swung; Oft their glad shouts, shrill and clear, Came upon the startled ear. Blended with the tremulous bleat, Of truant lambs, or voices sweet, Of birds, that take us by surprise, And mock the quickly-searching eyes. Nearer sat a fair-haired boy, Whistling with a thoughtless joy; A shepherd's crook was in his hand, Emblem of a mild command; And upon his rounded cheek Were hues that ripened apples streak. Disease, nor pain, nor sorrowing, Touched that small Arcadian king; His sinless subjects wandered free-- Confusion without anarchy. Happier he upon his throne. The breezy hill--though all alone-- Than the grandest monarchs proud Who mistrust the kneeling crowd. On a gently rising ground, The lovely valley's farthest bound, Bordered by an ancient wood, The cots in thicker clusters stood; And a church, uprose between, Hallowing the peaceful scene. Distance o'er its old walls threw A soft and dim cerulean hue, While the sun-lit gilded spire Gleamed as with celestial fire! I have crossed the ocean wave, Haply for a foreign grave; Haply never more to look On a British hill or brook; Haply never more to hear Sounds unto my childhood dear; Yet if sometimes on my soul Bitter thoughts beyond controul Throw a shade more dark than night, Soon upon the mental sight Flashes forth a pleasant ray Brighter, holier than the day; And unto that happy mood All seems beautiful and good.

D.L.R.

LINES TO A LADY,

WHO PRESENTED THE AUTHOR WITH SOME ENGLISH FRUITS AND FLOWERS.

Green herbs and gushing springs in some hot waste Though, grateful to the traveller's sight and taste, Seem far less sweet and fair than fruits and flowers That breathe, in foreign lands, of English bowers. Thy gracious gift, dear lady, well recalls Sweet scenes of home,--the white cot's trellised walls-- The trim red garden path--the rustic seat-- The jasmine-covered arbour, fit retreat For hearts that love repose. Each spot displays Some long-remembered charm. In sweet amaze I feel as one who from a weary dream Of exile wakes, and sees the morning beam Illume the glorious clouds of every hue That float o'er scenes his happy childhood knew. How small a spark may kindle fancy's flame And light up all the past! The very same Glad sounds and sights that charmed my heart of old Arrest me now--I hear them and behold. Ah! yonder is the happy circle seated Within, the favorite bower! I am greeted With joyous shouts; my rosy boys have heard A father's voice--their little hearts are stirred With eager hope of some new toy or treat And on they rush, with never-resting feet!

Gone is the sweet illusion--like a scene Formed by the western vapors, when between The dusky earth, and day's departing light The curtain falls of India's sudden night.

D.L.R.

The verdant carpet embroidered with little stars of gold and silver--the short-grown, smooth, and close-woven, but most delicate and elastic fresh sward--so soothing to the dazzled eye, so welcome to the wearied limbs--so suggestive of innocent and happy thoughts,--so refreshing to the freed visitor, long pent up in the smoky city--is surely no where to be seen in such exquisite perfection as on the broad meadows and softly- swelling hills of England. And perhaps in no country in the world could *pic-nic* holiday-makers or playful children with more perfect security of life and health stroll about or rest upon Earth's richly enamelled floor from sunrise to sunset on a summer's day. No Englishman would dare to stretch himself at full length and address himself to sleep upon an Oriental meadow unless he were perfectly indifferent to life itself and could see nothing terrible in the hostility of the deadliest reptiles. When wading through the long grass and thick jungles of Bengal, he is made to acknowledge the full force of the true and beautiful expression--"*In the midst of life we are in death.*" The British Indian exile on his return home is delighted with the "sweet security" of his native fields. He may then feel with Wordsworth how

> Dear is the forest frowning o'er his head. And dear *the velvet greensward* to his tread.

Or he may exclaim in the words of poor Keats--now slumbering under a foreign turf--

> Happy is England! I could be content To see no other verdure than her own.

It is a pleasing proof of the fine moral influence of natural scenery that the most ceremonious strangers can hardly be long seated together in the open air on the "velvet greensward" without casting off for a while the cold formalities of artificial life, and becoming as frank and social as ingenuous school-boys. Nature breathes peace and geniality into almost every human heart.

"John Thelwall," says Coleridge, "had something very good about him. We were sitting in a beautiful recess in the Quantocks when I said to him 'Citizen John, this is a fine place to talk treason in!' 'Nay, Citizen Samuel,' replied he, 'it is rather a place to make us forget that there is any necessity for treason!'"

Leigh Hunt, who always looks on nature with the eye of a true painter and the imagination of a true poet, has represented with delightful force and vividness some of those accidents of light and shade that diversify an English meadow.

RAIN AND SUNSHINE IN MAY.

"Can any thing be more lovely, than the meadows between the rains of May, when the sun smites them on the sudden like a painter, and they laugh up at him, as if he had lighted a loving cheek!

I speak of a season when the returning threats of cold and the resisting warmth of summer time, make robust mirth in the air; when the winds imitate on a sudden the vehemence of winter; and silver-white clouds are abrupt in their coming down and shadows on the grass chase one another, panting, over the fields, like a pursuit of spirits. With undulating necks they pant forward, like hounds or the leopard.

See! the cloud is after the light, gliding over the country like the shadow of a god; and now the meadows are lit up here and there with sunshine, as if the soul of Titian were standing in heaven, and playing his fancies on them. Green are the trees in shadow; but the trees in the sun how twenty-fold green *they* are--rich and variegated with gold!"

One of the many exquisite out-of-doors enjoyments for the observers of nature, is the sight of an English harvest. How cheering it is to behold the sickles flashing in the sun, as the reapers with well sinewed arm, and with a sweeping movement, mow down the close-arrayed ranks of the harvest field! What are "the rapture of the strife" and all the "pomp, pride and circumstance of glorious war," that bring death to some and agony and grief to others, compared with the green and golden trophies of the honest Husbandman whose bloodless blade makes no wife a widow, no child an orphan,--whose office is not to spread horror and desolation through shrieking cities, but to multiply and distribute the riches of nature over a smiling land.

But let us quit the open fields for a time, and turn again to the flowery retreats of

> Retired Leisure That in trim gardens takes his pleasure.

In all ages, in all countries, in all creeds, a garden is represented as the scene not only of earthly but of celestial enjoyment. The ancients had their Elysian Fields and the garden of the Hesperides, the Christian has his Garden of Eden, the Mahommedan his Paradise of groves and flowers and crystal fountains and black eyed Houries.

"God Almighty," says Lord Bacon, "first planted a garden; and indeed it is the purest of all pleasures: it is the greatest refreshment to the spirits of man." Bacon, though a utilitarian philosopher, was such a lover of flowers that he was never satisfied unless he saw them in almost every room of his house, and when he came to discourse of them in his Essays, his thoughts involuntarily moved harmonious numbers. How naturally the following prose sentence in Bacon's Essay on Gardens almost resolves itself into verse.

"For the heath which was the first part of our plot, I wish it to be framed as much as may be to a natural wildness. Trees I would have none in it, but some thickets made only of sweet briar and honeysuckle, and some wild vine amongst; and the ground set with violets, strawberries and primroses; for these are sweet, and prosper in the shade."

> "For the heath which was the third part of our plot-- I wish it to be framed As much as may be to a natural wildness. Trees I'd have none in't, but some thickets made Only of sweet-briar and honey-suckle, And some wild vine amongst; and the ground set With violets, strawberries, and primroses; For these are sweet and prosper in the shade."

It has been observed that the love of gardens is the only passion which increases with age. It is generally the most indulged in the two extremes of life. In middle age men are often too much involved in the affairs of the busy world fully to appreciate the tranquil pleasures in the gift of Flora. Flowers are the toys of the young and a source of the sweetest and serenest enjoyments for the old. But there is no season of life for which they are unfitted and of which they cannot increase the charm.

"Give me," says the poet Rogers, "a garden well kept, however small, two or three spreading trees and a mind at ease, and I defy the world." The poet adds that he would not have his garden, too much extended. He seems to think it possible to have too much of a good thing. "Three acres of flowers and a regiment of gardeners," he says, "bring no more pleasure than a sufficiency." "A hundred thousand roses," he adds, "which we look at *en masse*, do not identify themselves in the same manner as even a very small border; and hence, if the cottager's mind is properly attuned, the little cottage-garden may give him more real delight than belongs to the owner of a thousand acres." In a smaller garden "we become acquainted, as it were," says the same poet, "and even form friendships with, individual flowers." It is delightful to observe how nature thus adjusts the inequalities of fortune and puts the poor man, in point of innocent happiness, on a level with the rich. The man of the most moderate means may cultivate many elegant tastes, and may have flowers in his little garden that the greatest sovereign in the world might enthusiastically admire. Flowers are never vulgar. A rose from a peasant's patch of ground is as fresh and elegant and fragrant as if it had been nurtured in a Royal parterre, and it would not be out of place in the richest porcelain vase of the most aristocratical drawing-room in Europe. The poor man's flower is a present for a princess, and of all gifts it is the one least liable to be rejected even by the haughty. It might he worn on the fair brow or bosom of Queen Victoria with a nobler grace than the costliest or most elaborate production of the goldsmith or the milliner.

The majority of mankind, in the most active spheres of life, have moments in which they sigh for rural retirement, and seldom dream of such a retreat without making a garden the leading charm of it. Sir Henry Wotton says that Lord Bacon's garden was one of the best that he had seen either at home or abroad. Evelyn, the author of "Sylva, or a Discourse of Forest Trees," dwells with fond admiration, and a pleasing egotism, on the charms of his own beautiful and highly cultivated estate at Wooton in the county of Surrey. He tells us that the house is large and ancient and is "sweetly environed with delicious streams and venerable woods." "I will say nothing," he continues, "of the air, because the pre-eminence is universally given to Surrey, the soil being dry and sandy; but I should speak much of the gardens, fountains and groves that adorn it, were they not generally known to be amongst the most natural, and (till this later and universal luxury of the whole nation, since abounding in such expenses) the most magnificent that England afforded, and which indeed gave one of the first examples to that elegancy, since so much in vogue and followed, for the managing of their waters and other elegancies of that nature." Before he came into the possession of his paternal estate he resided at *Say's Court*, near Deptford, an estate which he possessed by purchase, and where he had a superb holly hedge four hundred feet long, nine feet high and five feet broad. Of this hedge, he was particularly proud, and he exultantly asks, "Is there under heaven a more glorious and refreshing object of the kind?" When the Czar of Muscovy visited England in 1698 to instruct himself in the art of ship-building, he had the use of Evelyn's house and garden, at *Say's Court*, and while there did so much damage to the latter that the owner loudly and bitterly complained. At last the Government gave Evelyn £150 as an indemnification. Czar Peter's favorite amusement was to ride in a wheel barrow through what its owner had once called the "impregnable hedge of holly." Evelyn was passionately fond of gardening. "The life and felicity of an excellent gardener," he observes, "is preferable to all other diversions." His faith in the art of Landscape-gardening was unwavering. It could *remove mountains*. Here is an extract from his Diary.

> "Gave his brother some directions about his garden" (at Wooton Surrey), "which, he was desirous to put into some form, for which he was to remove a mountain overgrown with large trees and thickets and a moat within ten yards of the house."

No sooner said than done. His brother dug down the mountain and "flinging it into a rapid stream (which carried away the sand) filled up the moat and levelled that noble area where now the garden and fountain is."

Though Evelyn dearly loved a garden, his chief delight was not in flowers but in forest trees, and he was more anxious to improve the growth of plants indigenous to the soil than to introduce exotics.[007]

Sir William Temple was so attached to his garden, that he left directions in his will that his heart should be buried there. It was enclosed in a silver box and placed under a sun-dial.

Dr. Thomson Reid, the eminent Scottish metaphysician, used to be found working in his garden in his eighty-seventh year.

The name of Chatham is in the long list of eminent men who have enjoyed a garden. We are told that "he loved the country: took peculiar pleasure in gardening; and had an extremely happy taste in laying out grounds." What a delightful thing it must have been for that great statesman, thus to relieve his mind from the weight of public care in the midst of quiet bowers planted and trained by his own hand!

Burton, in his *Anatomy of Melancholy*, notices the attractions of a garden as amongst the finest remedies for depression of the mind. I must give the following extracts from his quaint but interesting pages.

"To see the pleasant fields, the crystal fountains, And take the gentle air amongst the mountains.

"To walk amongst orchards, gardens, bowers, mounts, and arbours, artificial wildernesses, green thickets, arches, groves, lawns, rivulets, fountains, and such like pleasant places, (like that Antiochian Daphne,) brooks, pools, fishponds, between wood and water, in a fair meadow, by a river side, *ubi variae avium cantationes, florum colores, pratorum frutices,* &c. to disport in some pleasant plain, or park, run up a steep hill sometimes, or sit in a shady seat, must needs be a delectable recreation. *Hortus principis et domus ad delectationem facta, cum sylvâ, monte et piscinâ, vulgò la montagna*: the prince's garden at Ferrara, Schottus highly magnifies, with the groves, mountains, ponds, for a delectable prospect; he was much affected with it; a Persian paradise, or pleasant park, could not be more delectable in his sight. St. Bernard, in the description of his monastery, is almost ravished with the pleasures of it. "A sick man (saith he) sits upon a green bank, and when the dog-star parcheth the plains, and dries up rivers, he lies in a shady bower," *Fronde sub arborea ferventia temperat astra,* "and feeds his eyes with variety of objects, herbs, trees, to comfort his misery; he receives many delightsome smells, and fills his ears with that sweet and various harmony of birds; *good God,* (saith he), *what a company of pleasures hast thou made for man!*"

"The country hath his recreations, the city his several gymnics and exercises, May games, feasts, wakes, and merry meetings to solace themselves; the very being in the country; that life itself is a sufficient recreation to some men, to enjoy such pleasures, as those old patriarchs did. Dioclesian, the emperor, was so much affected with it, that he gave over his sceptre, and turned gardener. Constantine wrote twenty books of husbandry. Lysander, when ambassadors came to see him, bragged of nothing more than of his orchard, *hi sunt ordines mei*. What shall I say of Cincinnatus, Cato, Tully, and many such? how they have been pleased with it, to prune, plant, inoculate and graft, to show so many several kinds of pears, apples, plums, peaches, &c."

The Romans of all ranks made use of flowers as ornaments and emblems, but they were not generally so fond of directing or assisting the gardener, or taking the spade or hoe into their own hands, as are the British peasantry, gentry and nobility of the present day. They were not amateur Florists. They prized highly their fruit trees and pastures and cool grottoes and umbrageous groves; but they expended comparatively little time, skill or taste upon the flower-garden. Even their love of nature, though thoroughly genuine as far as it went, did not imply that minute and exact knowledge of her charms which characterizes some of our best British poets. They had no Thompson or Cowper. Their country seats were richer in architectural than floral beauty. Tully's Tuscan Villa, so fondly and minutely described by the proprietor himself, would appear to little advantage in the eyes of a true worshipper of Flora, if compared with Pope's retreat at Twickenham. The ancients had a taste for the *rural*, not for the *gardenesque*, nor perhaps even for the *picturesque*. The English have a taste for all three. Hence they have good landscape-gardeners and first-rate landscape-painters. The old Romans had neither. But though, some of our Spitalfields weavers have shown a deeper love, and perhaps even a finer taste, for flowers, than were exhibited by the citizens of Rome, abundant evidence is furnished to us by the poets in all ages and in all countries that nature, in some form or another has ever charmed the eye and the heart of man. The following version of a famous passage in Virgil, especially the lines in Italics, may give the English reader some idea of a Roman's dream of

RURAL HAPPINESS.

> Ah! happy Swains! if they their bliss but knew, Whom, far from boisterous war, Earth's bosom true With easy food supplies. If they behold No lofty dome its gorgeous gates unfold And pour at morn from all its chambers wide Of flattering visitants the mighty tide; Nor gaze on beauteous columns richly wrought, Or tissued robes, or busts from Corinth brought; Nor their white wool with Tyrian poison soil,

Nor taint with Cassia's bark their native oil; *Yet peace is theirs; a life true bliss that yields; And various wealth; leisure mid ample fields, Grottoes, and living lakes, and vallies green, And lowing herds; and 'neath a sylvan screen, Delicious slumbers. There the lawn and cave With beasts of chase abound.* The young ne'er crave A prouder lot; their patient toil is cheered; Their Gods are worshipped and their sires revered; And there when Justice passed from earth away She left the latest traces of her sway.

D.L.R.

Lord Bacon was perhaps the first Englishman who endeavored to reform the old system of English gardening, and to show that it was contrary to good taste and an insult to nature. "As for making knots or figures," he says, "with divers colored earths, that may lie under the windows of the house on that side on which the garden stands, they be but toys: you may see as good sights many times in tarts." Bacon here alludes, I suppose, to the old Dutch fashion of dividing flowerbeds into many compartments, and instead of filling them with flowers, covering one with red brick dust, another with charcoal, a third with yellow sand, a fourth with chalk, a fifth with broken China, and others with green glass, or with spars and ores. But Milton, in his exquisite description of the garden of Eden, does not allude to the same absurd fashion when he speaks of "curious knots,"

> Which not nice art, In beds and *curious knots*, but nature boon Poured forth profuse on hill and dale and plain.

By these *curious knots* the poet seems to allude, not to figures of "divers colored earth," but to the artificial and complicated arrangements and divisions of flowers and flower-beds.

Though Bacon went not quite so freely to nature as our latest landscape- gardeners have done, he made the *first step* in the right direction and deserves therefore the compliment which Mason has paid him in his poem of *The English Garden*.

> On thy realm Philosophy his sovereign lustre spread; Yet did he deign to light with casual glance The wilds of Taste, Yes, sagest Verulam, 'Twas thine to banish from the royal groves Each childish vanity of crisped knot[008] And sculptured foliage; to the lawn restore Its ample space, and bid it feast the sight With verdure pure, unbroken, unabridged; For verdure soothes the eye, as roseate sweets The smell, or music's melting strains the ear.

Yes--"*verdure soothes the eye*:"--and the mind too. Bacon himself observes, that "nothing is more pleasant to the eye than green grass kept finely shorn." Mason slightly qualifies his commendation of "the sage" by admitting that he had not quite completed his emancipation from the bad taste of his day.

> Witness his high arched hedge In pillored state by carpentry upborn, With colored mirrors decked and prisoned birds. But, when our step has paced the proud parterre, And reached the heath, then Nature glads our eye Sporting in all her lovely carelessness, There smiles in varied tufts the velvet rose, There flaunts the gadding woodbine, swells the ground In gentle hillocks, and around its sides Through blossomed shades the secret pathway steals.

The English Garden.

In one of the notes to *The English Garden* it is stated that "Bacon was the prophet, Milton the herald of modern Gardening; and Addison, Pope, and Kent the champions of true taste." Kent was by profession both a Painter and a Landscape-Gardener. Addison who had a pretty little retreat at Bilton, near Rugby, evinces in most of his occasional allusions to gardens a correct judgment. He complains that even in *his* time our British gardeners, instead of humouring nature, loved to deviate from it as much as possible. The system of verdant sculpture had not gone out of fashion. Our trees still rose in cones, globes, and pyramids. The work of the scissors was on every plant and bush. It was Pope, however, who did most to bring the topiary style into contempt and to encourage a more natural taste, by his humorous paper in the *Guardian* and his poetical

Epistle to the Earl of Burlington. Gray, the poet, observes in one of his letters, that "our skill in gardening, or rather laying out grounds, is the only taste we can call our own; the only proof of original talent in matters of pleasure. This is no small honor to us;" he continues, "since neither France nor Italy, has ever had the least notion of it." "Whatever may have been reported, whether truly or falsely" (says a contributor to *The World*) "of the Chinese gardens, it is certain that we are the first of the Europeans who have founded this taste; and we have been so fortunate in the genius of those who have had the direction of some of the finest spots of ground, that we may now boast a success equal to that profusion of expense which has been destined to promote the rapid progress of this happy enthusiasm. Our gardens are already the astonishment of foreigners, and, in proportion as they accustom themselves to consider and understand them will become their admiration." The periodical from which this is taken was published exactly a century ago, and the writer's prophecy has been long verified. Foreigners send to us for gardeners to help them to lay out their grounds in the English fashion. And we are told by the writer of an interesting article on gardens, in the *Quarterly Review*, that "the lawns at Paris, to say nothing of Naples, are regularly irrigated to keep up even the semblance of English verdure; and at the gardens of Versailles, and Caserta, near Naples, the walks have been supplied from the Kensington gravel-pits." "It is not probably known," adds the same writer, "that among our exportations every year is a large quantity of evergreens for the markets of France and Germany, and that there are some nurserymen almost wholly engaged in this branch of trade."

Pomfret, a poet of small powers, if a poet at all, has yet contrived to produce a popular composition in verse--*The Choice*--because he has touched with great good fortune on some of the sweetest domestic hopes and enjoyments of his countrymen.

> If Heaven the grateful liberty would give That I might choose my method how to live; And all those hours propitious Fate should lend In blissful ease and satisfaction spend; Near some fair town I'd have a private seat Built uniform; not little;

nor too great: Better if on a rising ground it stood, On this side fields, on that a neighbouring wood.

The Choice.

Pomfret perhaps illustrates the general taste when he places his garden "*near some fair town.*" Our present laureate, though a truly inspired poet, and a genuine lover of Nature even in her remotest retreats, has the garden of his preference, "*not quite beyond the busy world.*"

> Not wholly in the busy world, nor quite Beyond it, blooms the garden that I love, News from the humming city comes to it In sound of funeral or of marriage bells; And sitting muffled in dark leaves you hear The windy clanging of the minster clock; Although between it and the garden lies A league of grass.

Even "sounds inharmonious in themselves and harsh" are often pleasing when mellowed by the space of air through which they pass.

> 'Tis distance lends enchantment to the *sound*.

Shelley, in one of his sweetest poems, speaking of a scene in the neighbourhood of Naples, beautifully says:--

> Like many a voice of one delight, The winds, the birds, the ocean floods, *The city's voice itself is soft*, like solitude's.

No doubt the feeling that we are *near* the crowd but not *in* it, may deepen the sense of our own happy rural seclusion and doubly endear that pensive leisure in which we can "think down hours to moments," and in

> This our life, exempt from public haunt, Find tongues in trees, books in the running brooks, Sermons in stones, and good in every thing.

Shakespeare.

Besides, to speak truly, few men, however studious or philosophical, desire a total isolation from the world. It is pleasant to be able to take a sort of side glance at humanity, even when we are most in love with nature, and to feel that we can join our fellow creatures again when the social feeling returns upon us. Man was not made to live alone. Cowper, though he clearly loved retirement and a garden, did not desire to have the pleasure entirely to himself. "Grant me," he says, "a friend in my retreat."

> To whom to whisper solitude is sweet.

Cowper lived and died a bachelor. In the case of a married man and a father, garden delights are doubled by the presence of the family and friends, if wife and children happen to be what they should be, and the friends are genuine and genial.

All true poets delight in gardens. The truest that ever lived spent his latter days at New Place in Stratford-upon-Avon. He had a spacious and beautiful garden. Charles Knight tells us that "the Avon washed its banks; and within its enclosures it had its sunny terraces and green lawns, its pleached alleys and honeysuckle bowers," In this garden Shakespeare planted with his own hands his celebrated Mulberry tree. It was a noble specimen of the black Mulberry introduced into England in 1548[009]. In 1605, James I. issued a Royal edict recommending the cultivation of silkworms and offering packets of mulberry seeds to those amongst his subjects who were willing to sow them. Shakespeare's tree was planted in 1609. Mr. Loudon, observes that the black Mulberry has been known from the earliest records of antiquity and that it is twice mentioned in the Bible: namely, in the second Book of Samuel and in the Psalms. When New Place was in the possession of Sir Hough Clopton, who was proud of its interesting association with the history of our great poet, not only were Garrick and Macklin most hospitably entertained under the Mulberry tree, but all strangers on a proper application were admitted to a sight of it. But when Sir Hough Clopton was succeeded by the Reverend Francis Gastrell, that gentleman, to save himself the trouble of showing the tree to visitors, had "the gothic barbarity" to cut down and root up that interesting--indeed *sacred* memorial--of the Pride of the British Isles. The people of Stratford were so enraged at this sacrilege that they broke Mr. Gastrell's windows. That prosaic personage at last found the place too hot for him, and took his departure from a town whose inhabitants "doated on his very absence;" but before he went he completed the fall sum of his sins against good taste and good feeling by pulling to the ground the house in which Shakespeare had lived and died. This was done, it is said, out of sheer spite to the towns-people, with some of whom Mr. Gastrell had had a dispute about the rate at which the house was taxed. His change of residence was no great relief to him, for the whole British public felt sorely aggrieved, and wherever he went he was peppered with all sorts of squibs and satires. He "slid into verse," and "hitched in a rhyme."

> Sacred to ridicule his whole life long, And the sad burden of a merry song.

Thomas Sharp, a watchmaker, got possession of the fragments of Shakespeare's Mulberry tree, and worked them into all sorts of elegant ornaments and toys, and disposed of them at great prices. The corporation of Stratford presented Garrick with the freedom of the town in a box made of the wood of this famous tree, and the compliment seems to have suggested to him his public festival or pageant in honor of the poet. This Jubilee, which was got up with great zeal, and at great expense and trouble, was attended by vast throngs of the admirers of Shakespeare from all parts of the kingdom. It was repeated on the stage and became so popular as a theatrical exhibition that it was represented night after night for more than half a season to crowded audiences.

Upon the subject of gardens, let us hear what has been said by the self- styled "melancholy Cowley." When in the smoky city pent, amidst the busy hum of men, he sighed unceasingly for some green retreat. As he paced the crowded thorough-fares of London, he thought of the velvet turf and the pure air of the country. His imagination carried him into secluded groves or to the bank of a murmuring river, or into some trim and quiet garden. "I never," he says, "had any other desire so strong and so like to covetousness, as that one which I have had always, that I might be master at last of a small house and a large garden, with very moderate conveniences joined to them, and there dedicate the remainder of my life only to the culture of them and the study of nature," The late Miss Mitford, whose writings breathe so freshly of the nature that she loved so dearly, realized for herself a similar desire. It is said that she had the cottage of a peasant with the garden of a Duchess. Cowley is not contented with expressing in plain prose his appreciation of garden enjoyments. He repeatedly alludes to them in verse.

> Thus, thus (and this deserved great Virgil's praise) The old Corycian yeoman passed his days; Thus his wise life Abdolonymus spent; Th' ambassadors, which the great emperor sent To offer him a crown, with wonder found The reverend gardener, hoeing of his ground; Unwillingly and slow and discontent From his loved cottage to a throne he went; And oft he stopped, on his triumphant way: And oft looked back: and oft was heard to say Not without sighs, Alas! I there forsake A happier kingdom than I go to take.

Lib. IV. Plantarum.

Here is a similar allusion by the same poet to the delights which great men amongst the ancients have taken in a rural retirement.

> Methinks, I see great Dioclesian walk In the Salonian garden's noble shade Which by his own imperial hands was made, I see him smile, methinks, as he does talk With the ambassadors, who come in vain To entice him to a throne again. "If I, my friends," said he, "should to you show All the delights which in these gardens grow, 'Tis likelier much that you should with me stay, Than 'tis that you should carry me away: And trust me not, my friends, if every day I walk not here with more delight, Than ever, after the most happy sight In triumph to the Capitol I rode, To thank the gods, and to be thought myself almost a god,"

The Garden.

Cowley does not omit the important moral which a garden furnishes.

Where does the wisdom and the power divine In a more bright and sweet reflection shine? Where do we finer strokes and colors see Of the Creator's real poetry. Than when we with attention look Upon the third day's volume of the book? If we could open and intend our eye *We all, like Moses, might espy, E'en in a bush, the radiant Deity.*

In Leigh Hunt's charming book entitled *The Town*, I find the following notice of the partiality of poets for houses with gardens attached to them:--

"It is not surprizing that *garden-houses* as they were called; should have formerly abounded in Holborn, in Bunhill Row, and other (at that time) suburban places. We notice the fact, in order to observe *how fond the poets were of occupying houses of this description. Milton seems to have made a point of having one.* The only London residence of Chapman which is known, was in Old Street Road; doubtless at that time a rural suburb. Beaumont and Fletcher's house, on the Surrey side of the Thames, (for they lived as well as wrote together,) most probably had a garden; and Dryden's house in Gerard Street looked into the garden of the mansion built by the Earls of Leicester. A tree, or even a flower, put in a window in the streets of a great city, (and the London citizens, to their credit, are fond of flowers,) affects the eye something in the same way as the hand-organs, which bring unexpected music to the ear. They refresh the common-places of life, shed a harmony through the busy discord, and appeal to those first sources of emotion, which are associated with the remembrance of all that is young and innocent."

Milton must have been a passionate lover of flowers and flower-gardens or he could never have exhibited the exquisite taste and genial feeling which characterize all the floral allusions and descriptions with which so much of his poetry is embellished. He lived for some time in a house in Westminster over-looking the Park. The same house was tenanted by Jeremy Bentham for forty years. It would be difficult to meet with any two individuals of more opposite temperaments than the author of *Paradise Lost* and the Utilitarian Philosopher. There is or was a stone in the wall at the end of the garden inscribed TO THE PRINCE OF POETS. Two beautiful cotton trees overarched the inscription, "and to show" says Hazlitt, (who subsequently lived in the same house himself,) "how little the refinements of taste or fancy entered Bentham's system, he proposed at one time to cut down these beautiful trees, to convert the garden, where he had breathed an air of truth and heaven for near half a century, into a paltry Chreistomathic School, and to make Milton's house (the cradle of *Paradise Lost*) a thoroughfare, like a three-stalled stable, for the idle rabble of Westminster to pass backwards and forwards to it with their cloven hoofs!"

No poet, ancient or modern, has described a garden on a large scale in so noble a style as Milton. He has anticipated the finest conceptions of the latest landscape-gardeners, and infinitely surpassed all the accounts we have met with of the gardens of the olden time before us. His Paradise is a

> Spot more delicious than those gardens feigned Or of revived Adonis or renowned Alcinous, host of old Laertes' son Or that, not mystic, where the sapient King Held dalliance with his fair Egyptian spouse[010]

The description is too long to quote entire, but I must make room for a delightful extract. Familiar as it must be to all lovers of poetry, who will object to read it again and again? Genuine poetry is like a masterpiece of the painter's art:--we can gaze with admiration for the hundredth time on a noble picture. The mind and the eye are never satiated with the truly beautiful. "A thing of beauty is a joy for ever."

PARADISE.[011]

> So on he fares, and to the border comes Of Eden, where delicious Paradise, Now nearer, crowns with her enclosure green, As with a rural mound, the champaign head Of a steep wilderness, whose hairy sides With thicket overgrown, grotesque and wild, Access denied: and overhead up grew Insuperable height of loftiest shade, Cedar, and pine, and fir, and branching palm, A sylvan scene; and as, the ranks ascend Shade above shade, a woody theatre Of stateliest view. Yet higher than their tops, The verdurous wall of Paradise up-sprung: Which to our general sire gave prospect lar-

ge Into his nether empire neighbouring round; And higher than that wall a circling row Of goodliest trees, loaden with fairest fruit, Blossoms and fruits at once, of golden hue, Appear'd, with gay enamell'd colours mix'd; On which the sun more glad impress'd his beams, Than on fair evening cloud, or humid bow. When God hath shower'd the earth; so lovely seem'd That landscape: and of pure now purer air Meets his approach, and to the heart inspires Vernal delight and joy, able to drive All sadness but despair: now gentle gales, Fanning their odoriferous wings, dispense Native perfumes and whisper whence they stole Those balmy spoils. As when to them who sail Beyond the Cape of Hope, and now are past Mozambic, off at sea north-east winds blow Sabean odours from the spicy shore Of Araby the Blest; with such delay Well pleased they slack their course, and many a league Cheer'd with the grateful smell, old Ocean smiles.

Southward through Eden went a river large, Nor changed his course, but through the shaggy hill Pass'd underneath ingulf'd; for God had thrown That mountain as his garden mould, high raised Upon the rapid current, which through veins Of porous earth with kindly thirst up-drawn, Rose a fresh fountain, and with many a rill Water'd the garden; thence united fell Down the steep glade, and met the nether flood, Which from his darksome passage now appears; And now, divided into four main streams, Runs diverse, wandering many a famous realm And country, whereof here needs no account; But rather to tell how, if art could tell, How from that sapphire fount the crisped brooks, Rolling on orient pearl and sands of gold, With mazy error under pendent shades, Ran nectar, visiting each plant, and fed Flowers worthy of Paradise, which not nice art In beds and curious knots, but nature boon Pour'd forth profuse on hill, and dale, and plain, Both where the morning sun first warmly smote The open field, and where the unpierced shade Imbrown'd the noontide bowers; thus was this place A happy rural seat of various view; Groves whose rich, trees wept odorous gums and balm; Others whose fruit, burnish'd with golden rind, Hung amiable, Hesperian fables true, If true, here only, and of delicious taste:

> Betwixt them lawns, or level downs, and flocks Grazing the tender herb, were interposed; Or palmy hillock, or the flowery lap Of some irriguous valley spread her store, Flowers of all hue, and without thorn the rose: Another side, umbrageous grots and caves Of cool recess, o'er which the mantling vine Lays forth her purple grape, and gently creeps Luxuriant; meanwhile murmuring waters fall Down the slope hills, dispersed, or in a lake, That to the fringed bank with myrtle crown'd Her crystal mirror holds, unite their streams. The birds their quire apply; airs, vernal airs, Breathing the smell of field and grove attune, The trembling leaves, while universal Pan, Knit with the Graces and the Hours in dance, Led on the eternal Spring.

Pope in his grounds at Twickenham, and Shenstone in his garden farm of the Leasowes, taught their countrymen to understand how much taste and refinement of soul may be connected with the laying out of gardens and the cultivation of flowers. I am sorry to learn that the famous retreats of these poets are not now what they were. The lovely nest of the little Nightingale of Twickenham has fallen into vulgar hands. And when Mr. Loudon visited (in 1831) the once beautiful grounds of Shenstone, he "found them in a state of indescribable neglect and ruin."

Pope said that of all his works that of which he was proudest was his garden. It was of but five acres, or perhaps less, but to this he is said to have given a charming variety. He enumerates amongst the friends who assisted him in the improvement of his grounds, the gallant Earl of Peterborough "whose lightnings pierced the Iberian lines."

> Know, all the distant din that world can keep, Rolls o'er my grotto, and but soothes my sleep. There my retreat the best companions grace Chiefs out of war and statesmen out of place. There St. John mingles with my friendly bowl The feast of reason and the flow of soul; And he whose lightnings pierced the Iberian lines Now forms my quincunx and now ranks my vines; Or tames the genius of the stubborn plain Almost as quickly as he conquered Spain.

Frederick Prince of Wales took a lively interest in Pope's tasteful Tusculanum and made him a present of some urns or vases either for his "laurel circus or to terminate his points." His famous grotto, which he is so fond of alluding to, was excavated to avoid an inconvenience. His property lying on both sides of the public highway, he contrived his highly ornamented passage under the road to preserve privacy and to connect the two portions of his estate.

The poet has given us in one of his letters a long and lively description of his subterranean embellishments. But his verse will live longer than his prose. He has immortalized this grotto, so radiant with spars and ores and shells, in the following poetical inscription:--

> Thou, who shalt stop, where Thames' translucent wave Shines a broad mirror through the shadowy cave, Where lingering drops from mineral roofs distil, And pointed crystals break the sparkling rill, Unpolished gems no ray on pride bestow, And latent metals innocently glow, Approach! Great Nature studiously behold, And eye the mine without a wish for gold Approach--but awful! Lo, the Egerian grot, Where, nobly pensive, ST JOHN sat and thought, Where British sighs from dying WYNDHAM stole, And the bright flame was shot thro' MARCHMONT'S soul; Let such, such only, tread this sacred floor Who dare to love their country, and be poor.

Horace Walpole, speaking of the poet's garden, tells us that "the passing through the gloom from the grotto to the opening day, the retiring and again assembling shades, the dusky groves, the larger lawn, and the solemnity at the cypresses that led up to his mother's tomb, were managed with exquisite judgment."

> Cliveden's proud alcove, The bower of wanton Shrewsbury and love,

alluded to by Pope in his sketch of the character of Villiers, Duke of Buckingham, though laid out by Kent, was probably improved by the poet's suggestions. Walpole seems to think that the beautiful grounds at Rousham, laid out for General Dormer, were planned on the model of the garden at Twickenham, at least the opening and retiring "shades of Venus's Vale." And these grounds at Rousham were pronounced "the most engaging of all Kent's works." It is said that the design of the garden at Carlton House, was borrowed from that of Pope.

Wordsworth was correct in his observation that "Landscape gardening is a liberal art akin to the arts of poetry and painting." Walpole describes it as "an art that realizes painting and improves nature." "Mahomet," he adds, "imagined an Elysium, but Kent created many."

Pope's mansion was not a very spacious one, but it was large enough for a private gentleman of inexpensive habits. After the poet's death it was purchased by Sir William Stanhope who enlarged both the house and garden.[012] A bust of Pope, in white marble, has been placed over an arched way with the following inscription from the pen of Lord Nugent:

> The humble roof, the garden's scanty line, Ill suit the genius of the bard divine; But fancy now displays a fairer scope And Stanhope's plans unfold the soul of Pope.

I have not heard who set up this bust with its impudent inscription. I hope it was not Stanhope himself. I cannot help thinking that it would have been a truer compliment to the memory of Pope if the house and grounds had been kept up exactly as he had left them. Most people, I suspect, would greatly have preferred the poet's own "unfolding of his soul" to that "*unfolding*" attempted for him by a Stanhope and commemorated by a Nugent. Pope exhibited as much taste in laying out his grounds as in constructing his poems. Sir William, after his attempt to make the garden more worthy of the original designer, might just as modestly have undertaken to enlarge and improve the poetry of Pope on the plea that it did not sufficiently *unfold his soul*. A line of Lord Nugent's might in that case have been transferred from the marble bust to the printed volume:

> His fancy now displays a fairer scope.

Or the enlarger and improver might have taken his motto from Shakespeare:

> To my *unfolding* lend a gracious ear.

This would have been an appropriate motto for the title-page of "*The Poems of Pope: enlarged and improved: or The Soul of the Poet Unfolded.*"

But in sober truth, Pope, whether as a gardener or as a poet, required no enlarger or improver of his works. After Sir William Stanhope had left Pope's villa it came into the possession of Lord Mendip, who exhibited a proper respect for the poet's memory; but when in 1807 it was sold to the Baroness Howe, that lady pulled down the house and built another. The place subsequently came into the possession of a Mr. Young. The grounds have now no resemblance to what the taste of Pope had once made them. Even his mother's monument has been removed! Few things would have more deeply touched the heart of the poet than the anticipation of this insult to the memory of so revered a parent. His filial piety was as remarkable as his poetical genius. No passages in his works do him more honor both as a man and as a poet than those which are mellowed into a deeper tenderness of sentiment and a softer and sweeter music by his domestic affections. There are probably few readers of English poetry who have not the following lines by heart,

> Me, let the tender office long engage To rock the cradle of reposing age; With lenient arts extend a mother's breath; Make langour smile, and smooth the bed of death; Explore the thought, explain the asking eye, And keep at least one parent from the sky.

In a letter to Swift (dated March 29, 1731) begun by Lord Bolingbroke and concluded by Pope, the latter speaks thus touchingly of his dear old parent:

"My Lord has spoken justly of his lady; why not I of my mother? Yesterday was her birth-day, now entering on the ninety-first year of her age; her memory much diminished, but her senses very little hurt, her sight and hearing good; she sleeps not ill, eats moderately, drinks water, says her prayers; this is all she does. I have reason to thank God for continuing so long to me a very good and tender parent, and for allowing me to exercise for some years those cares which are now as necessary to her, as hers have been to me."

Pope lost his mother two years, two months, and a few days after the date of this letter. Three days after her death he entreated Richardson, the painter, to take a sketch of her face, as she lay in her coffin: and for this purpose Pope somewhat delayed her interment. "I thank God," he says, "her death was as easy as her life was innocent; and as it cost her not a groan, nor even a sigh, there is yet upon her countenance such an expression of tranquillity, nay almost of pleasure, that it is even amiable to behold it. It would afford the finest image of a saint expired, that ever painting drew, and it would be the greatest obligation which even that obliging art could ever bestow upon a friend if you would come and sketch it for me." The writer adds, "I shall hope to see you this evening, as late as you will, or to-morrow morning as early, *before this winter flower is faded*."

On the small obelisk in the garden, erected by Pope to the memory of his mother, he placed the following simple and pathetic inscription.

> AH! EDITHA! MATRUM OPTIMA! MULIERUM AMANTISSIMA! VALE!

I wonder that any one could have had the heart to remove or to destroy so interesting a memorial.

It is said that Pope planted his celebrated weeping willow at Twickenham with his own hands, and that it was the first of its particular species introduced into England. Happening to be with Lady Suffolk when she received a parcel from Spain, he observed that it was bound with green twigs which looked as if they might vegetate. "Perhaps," said he, "these may produce something that we have not yet in England." He tried a cutting, and it succeeded. The tree was removed by some person as barbarous as the reverend gentleman who cut down Shakespeare's Mulberry Tree. The Willow was destroyed for the same reason, as the Mulberry Tree--because the owner was annoyed at persons asking to see it. The Weeping Willow

> That shows his hoar leaves in the glassy stream,[013]

has had its interest with people in general much increased by its association with the history of Napoleon in the Island of St. Helena. The tree whose boughs seemed to hang so fondly over his remains has now its scions in all parts of the world. Few travellers visited the tomb without taking a small cutting of the Napoleon Willow for cultivation in their own land. Slips of the Willow at Twickenham, like those of the Willow at St. Helena, have also found their way into many countries. In 1789 the Empress of Russia had some of them planted in her garden at St. Petersburgh.

Mr. Loudon tells us that there is an old *oak* in Binfield Wood, Windsor Forest, which is called *Pope's Oak*, and which bears the inscription "HERE POPE SANG:"[014] but according to general tradition it was a *beech* tree, under which Pope wrote his "Windsor Forest." It is said that as that tree was decayed, Lady Gower had the inscription alluded to carved upon another tree near it. Perhaps the substituted tree was an oak.

I may here mention that in the Vale of Avoca there is a tree celebrated as that under which Thomas Moore wrote the verses entitled "The meeting of the Waters."

The allusion to *Pope's Oak* reminds me that Chaucer is said to have planted three oak trees in Donnington Park near Newbury. Not one of them is now, I believe, in existence. There is an oak tree in Windsor Forest above 1000 years old. In the hollow of this tree twenty people might be accommodated with standing room. It is called *King's Oak*: it was William the Conqueror's favorite tree. *Herne's Oak* in Windsor Park, is said by some to be still standing, but it is described as a mere anatomy.

> ----An old oak whose boughs are mossed with age, And high top bald with dry antiquity.

As You Like it.

"It stretches out its bare and sapless branches," says Mr. Jesse, "like the skeleton arms of some enormous giant, and is almost fearful in its decay." *Herne's Oak*, as every one knows, is immortalised by Shakespeare, who has spread its fame over many lands.

> There is an old tale goes that Herne the hunter, Sometime a keeper here in Windsor Forest, Doth all the winter time, at still midnight, Walk round about an oak, with great ragg'd horns, And there he blasts the trees, and takes the cattle; And makes milch cows yield blood, and shakes a chain In a most hideous and dreadful manner. You have heard of such a spirit; and well you know, The superstitious, idle-headed eld Received, and did deliver to our age, This tale of Herne the Hunter for a truth.

Merry Wives of Windsor.

"Herne, the hunter" is said to have hung himself upon one of the branches of this tree, and even,

> ----Yet there want not many that do fear, In deep of night to walk by this Herne's Oak.

Merry Wives of Windsor.

It was not long ago visited by the King of Prussia to whom Shakespeare had rendered it an object of great interest.

It is unpleasant to add that there is considerable doubt and dispute as to its identity. Charles Knight and a Quarterly Reviewer both maintain that *Herne's Oak* was cut down with a number of other old trees in obedience to an order from George the Third when he was not in his right mind, and that his Majesty deeply regretted the order he had given when he found that the most interesting tree in his Park had been destroyed. Mr. Jesse, in his *Gleanings in Natural History*, says that after some pains to ascertain the truth, he is convinced that this story is not correct, and that the famous old tree is still standing. He adds that George the Fourth often alluded to the story and said that though one of the trees cut down was supposed to have been *Herne's Oak*, it was not so in reality. George the Third, it is said, once called the attention of Mr. Ingalt, the manager of Windsor Home Park to a particular tree, and said "I brought you here to point out this tree to you. I commit it to your especial charge; and take care that no damage is ever done to it. I had rather that every tree in the park should be cut down than that this tree should be hurt. *This is Hernes Oak."*

Sir Philip Sidney's Oak at Penshurst mentioned by Ben Jonson--

> That taller tree, of which the nut was set At his great birth, where all the Muses met--

is still in existence. It is thirty feet in circumference. Waller also alludes to

> Yonder tree which stands the sacred mark Of noble Sidney's birth.

Yardley Oak, immortalized by Cowper, is now in a state of decay.

> Time made thee what thou wert--king of the woods! And time hath made thee what thou art--a cave For owls to roost in.

Cowper.

The tree is said to be at least fifteen hundred years old. It cannot hold its present place much longer; but for many centuries to come it will

> Live in description and look green in song.

It stands on the grounds of the Marquis of Northampton; and to prevent people from cutting off and carrying away pieces of it as relics, the following notice has been painted on a board and nailed to the tree:--"*Out of respect to the memory of the poet Cowper, the Marquis of Northampton is particularly desirous of preserving this Oak.*"

Lord Byron, in early life, planted an oak in the garden at Newstead and indulged the fancy, that as that flourished so should he. The oak has survived the poet, but it will not outlive the memory of its planter or even the boyish verses which he addressed to it.

Pope observes, that "a tree is a nobler object than a prince in his coronation robes." Yet probably the poet had never seen any tree larger than a British oak. What would he have thought of the Baobab tree in Abyssinia, which measures from 80 to 120 feet in girth, and sometimes reaches the age of five thousand years. We have no such sylvan patriarch in Europe. The oldest British tree I have heard of, is a yew tree of Fortingall in Scotland, of which the age is said to be two thousand five hundred years. If trees had long memories and could converse with man, what interesting chapters these survivors of centuries might add to the history of the world!

Pope was not always happy in his Twickenham Paradise. His rural delights were interrupted for a time by an unrequited passion for the beautiful and highly-gifted but eccentric Lady Mary Wortley Montague.

> Ah! friend, 'tis true--this truth you lovers know; In vain my structures rise, my gardens grow; In vain fair Thames reflects the double scenes Of hanging mountains and of sloping greens; Joy lives not here, to happier seats it flies, And only dwells where Wortley casts her eyes. What are the gay parterre, the chequered shade, The morning bower, the evening colonnade, But soft recesses of uneasy minds, To sigh unheard in to the passing winds? So the struck deer, in some sequestered part, Lies down to die, the arrow at his heart; He, stretched unseen, in coverts hid from day, Bleeds drop by drop, and pants his life away.

These are exquisite lines, and have given delight to innumerable readers, but they gave no delight to Lady Mary. In writing to her sister, the Countess of Mar, then at Paris, she says in allusion to these "most musical, most melancholy" verses--"*I stifled them here; and I beg they may die the same death at Paris.*" It is not, however, quite so easy a thing as Lady Mary seemed to think, to "stifle" such poetry as Pope's.

Pope's notions respecting the laying out of gardens are well expressed in the following extract from the fourth Epistle of his Moral Essays.[015] This fourth Epistle was addressed, as most readers will remember, to the accomplished Lord Burlington, who, as Walpole says, "had every quality of a genius and an artist, except envy. Though his own designs were more chaste and classic than Kent's, he entertained him in his house till his death, and was more studious to extend his friend's fame than his own."

> Something there is more needful than expense, And something previous e'en to taste--'tis sense; Good sense, which only is the gift of Heaven, And though no science fairly worth the seven; A light, which in yourself you must perceive; Jones and Le Nôtre have it not to give. To build, or plant, whatever you intend, To rear the column or the arch to bend; To swell the terrace, or to sink the grot; In all let Nature never be forgot. But treat the goddess like a modest fair, Nor over dress nor leave her wholly bare; Let not each beauty every where be spied, Where half the skill is decently to hide. He gains all points, who pleasingly confounds, Surprizes, varies, and conceals the bounds. *Consult the genius of the place in all*;[016] That tells the waters or to rise or fall; Or helps the ambitious hill the heavens to scale, Or scoops in circling theatres the vale; Calls in the country, catches opening glades, Joins willing woods and varies shades from shades; Now breaks, or now directs, th' intending lines; Paints as you plant, and, as you work, designs. Still follow sense, of every art the soul; Parts answering parts shall slide into a whole, Spontaneous beauties all around advance, Start e'en from difficulty, strike from chance; Nature shall join you; time shall make it grow A work to wonder at--perhaps a STOWE.[017] Without it proud Versailles![018] Thy glory falls; And Nero's terraces desert their walls. The vast parterres a thousand hands shall make, Lo! Cobham comes and floats them with a lake; Or cut wide views through mountains to the plain, You'll wish your hill or sheltered seat again.

Pope is in most instances singularly happy in his compliments, but the allusion to STOWE--as "*a work to wonder at*"--has rather an equivocal appearance, and so also has the mention of Lord Cobham, the proprietor of the place. In the first draught of the poem, the name of Bridgeman was inserted where Cobham's now stands, but as Bridgeman mistook the compliment for a sneer, the poet thought the landscape-gardener had proved himself undeserving of the intended honor, and presented the second-hand compliment to the peer. The grounds at Stowe, more praised by poets than any other private estate in England, extend to 400 acres. There are many other fine estates in our country of far greater extent, but of less celebrity. Some of them are much too extensive, perhaps, for true enjoyment. The Earl of Leicester, when he had completed his seat at Holkham, observed, that "It was a melancholy thing to stand alone in one's country. I look round; not a house is to be seen but mine. I am the Giant of Giant-castle and have ate up all my neighbours." The Earl must have felt that the political economy of Goldsmith in his *Deserted Village* was not wholly the work of imagination.

> Sweet smiling village! Loveliest of the lawn, Thy sports are fled and all thy charms withdrawn; Amidst thy bowers the tyrant's hand is seen And desolation saddens all the green,-- *One only master grasps thy whole domain.*
>
> Where then, ah! where shall poverty reside, To scape the pressure of contiguous pride?

"Hearty, cheerful Mr. Cotton," as Lamb calls him, describes Stowe as a Paradise.

ON LORD COBHAM'S GARDEN.

> It puzzles much the sage's brains Where Eden stood of yore, Some place it in Arabia's plains, Some say it is no more. But Cobham can these tales confute, As all the curious know; For he hath proved beyond dispute, That Paradise is STOWE.

Thomson also calls the place a paradise:

> Ye Powers That o'er the garden and the rural seat Preside, which shining through the cheerful land In countless numbers blest Britannia sees; O, lead me to the wide-extended walks, *The fair majestic paradise of Stowe!* Not Persian Cyrus on Ionia's shore E'er saw such sylvan scenes; such various art By

genius fired, such ardent genius tamed By cool judicious art, that in the strife All-beauteous Nature fears to be out-done.

The poet somewhat mars the effect of this compliment to the charms of Stowe, by making it a matter of regret that the owner

His verdant files Of ordered trees should here inglorious range, Instead of squadrons flaming o'er the field, And long embattled hosts.

This representation of rural pursuits as inglorious, a sentiment so out of keeping with his subject, is soon after followed rather inconsistently, by a sort of paraphrase of Virgil's celebrated picture of rural felicity, and some of Thomson's own thoughts on the advantages of a retreat from active life.

Oh, knew he but his happiness, of men The happiest he! Who far from public rage Deep in the vale, with a choice few retired Drinks the pure pleasures of the rural life, &c.

Then again:--

Let others brave the flood in quest of gain And beat for joyless months, the gloomy wave. *Let such as deem it glory to destroy, Rush into blood, the sack of cities seek; Unpierced, exulting in the widow's wail, The virgin's shriek and infant's trembling cry.*

While he, from all the stormy passions free That restless men involve, hears and *but* hears, At distance safe, the human tempest roar, Wrapt close in conscious peace. The fall of kings, The rage of nations, and the crush of states, Move not the man, who from the world escaped, In still retreats and flowery solitudes, To nature's voice attends, from month to month, And day to day, through the revolving year; Admiring sees her in her every shape; Feels all her sweet emotions at his heart; Takes what she liberal gives, nor asks for more. He, when young Spring, protudes the bursting gems Marks the first bud, and sucks the healthful gale Into his freshened soul; her genial hour He full enjoys, and not a beauty blows And not an opening blossom breathes in vain.

Thomson in his description of Lord Townshend's seat of Rainham--another English estate once much celebrated and still much admired--exclaims:

> Such are thy beauties, Rainham, such the haunts Of angels, in primeval guiltless days When man, imparadised, conversed with God.

And Broome after quoting the whole description in his dedication of his own poems to Lord Townshend, observes, in the old fashioned fulsome strain, "This, my lord, is but a faint picture of the place of your retirement which no one ever enjoyed more elegantly."[019] "A faint picture!" What more would the dedicator have wished Thomson to say? Broome, if not contented with his patron's seat being described as an earthly Paradise, must have desired it to be compared with Heaven itself, and thus have left his Lordship no hope of the enjoyment of a better place than he already possessed.

Samuel Boyse, who when without a shirt to his back sat up in his bed to write verses, with his arms through two holes in his blanket, and when he went into the streets wore paper collars to conceal the sad deficiency of linen, has a poem of considerable length entitled *The Triumphs of Nature*. It is wholly devoted to a description of this magnificent garden,[020] in which, amongst other architectural ornaments, was a temple dedicated to British worthies, where the busts of Pope and Congreve held conspicuous places. I may as well give a specimen of the lines of poor Boyse. Here is his description of that part of Lord Cobham's grounds in which is erected to the Goddess of Love, a Temple containing a statue of the Venus de Medicis.

> Next to the fair ascent our steps we traced, Where shines afar the bold rotunda placed; The artful dome Ionic columns bear Light as the fabric swells in ambient air. Beneath enshrined the Tuscan Venus stands And beauty's queen the beauteous scene commands: The fond beholder sees with glad surprize, Streams glisten, lawns appear, and forests rise-- Here

through thick shades alternate buildings break, There through the borders steals the silver lake, A soft variety delights the soul, And harmony resulting crowns the whole.

Congreve in his Letter in verse addressed to Lord Cobham asks him to

Tell how his pleasing Stowe employs his time.

It would seem that the proprietor of Stowe took particular interest in the disposition of the water on his grounds. Congreve enquires

Or dost thou give the winds afar to blow Each vexing thought, and heart-devouring woe, And fix thy mind alone on rural scenes, *To turn the level lawns to liquid plains*? To raise the creeping rills from humble beds And force the latent spring to lift their heads, On watery columns, capitals to rear, That mix their flowing curls with upper air?

Or slowly walk along the mazy wood To meditate on all that's wise and good.

The line:--

To turn the level lawn to liquid plains--

Will remind the reader of Pope's

Lo! Cobham comes and floats them with a lake--

And it might be thought that Congreve had taken the hint from the bard of Twickenham if Congreve's poem had not preceded that of Pope. The one was published in 1729, the other in 1731.

Cowper is in the list of poets who have alluded to "Cobham's groves" and Pope's commemoration of them.

And *Cobham's groves* and Windsor's green retreats When Pope describes them have a thousand sweets.

"Magnificence and splendour," says Mr. Whately, the author of *Observations on Modern Gardening*, "are the characteristics of Stowe. It is like one of those places celebrated in antiquity which were devoted to the purposes of religion, and filled with sacred groves, hallowed fountains, and temples dedicated to several deities; the resort of distant nations and the object of veneration to half the heathen world: the pomp is, at Stowe, blended with beauty; and the place is equally distinguished by its amenity and grandeur." Horace Walpole speaks of its "visionary enchantment." "I have been strolling about in Buckinghamshire and Oxfordshire, from garden to garden," says Pope in one of his letters, "but still returning to Lord Cobham's with fresh satisfaction."[021]

The grounds at Stowe, until the year 1714, were laid out in the old formal style. Bridgeman then commenced the improvements and Kent subsequently completed them.

Stowe is now, I believe, in the possession of the Marquis of Chandos, son of the Duke of Buckingham. It is melancholy to state that the library, the statues, the furniture, and even some of the timber on the estate, were sold in 1848 to satisfy the creditors of the Duke.

Pope was never tired of improving his own grounds. "I pity you, Sir," said a friend to him, "because you have now completed every thing belonging to your gardens."[022] "Why," replied Pope, "I really shall be at a loss for the diversion I used to take in carrying out and finishing things: I have now nothing left me to do but to add a little ornament or two along the line of the Thames." I dare say Pope was by no means so near the end of his improvements as he and his friend imagined. One little change in a garden is sure to suggest or be followed by another. Garden-improvements are "never ending, still beginning." The late Dr. Arnold, the famous schoolmaster, writing to a friend, says--"The garden is a constant source of amusement to us both (self and wife); there are always some little alterations to be made, some few spots where an additional shrub or two would be ornamental, something coming into blossom; so that I can always delight to go round and see how things are going on." A garden is indeed a scene of continual change. Nature, even without the aid of the gardener, has "infinite variety," and supplies "a perpetual feast of nectared sweets where no crude surfeit reigns."

Spence reports Pope to have said: "I have sometimes had an idea of planting an old gothic cathedral in trees. Good large poplars, with their white stems, cleared of boughs to a proper height would serve very well for the columns, and might form the different aisles or peristilliums, by their different distances and heights. These would look very well near, and the dome rising all in a proper tuft in the middle would look well at a distance." This sort of verdant architecture would perhaps have a pleasing effect, but it is rather too much in the artificial style, to be quite consistent with Pope's own idea of landscape-gardening. And there are other trees that would form a nobler natural cathedral than the formal poplar. Cowper did not think of the poplar, when he described a green temple-roof.

> How airy and how light the graceful arch, Yet awful as the consecrated roof Re-echoing pious anthems.

Almost the only traces of Pope's garden that now remain are the splendid Spanish chesnut-trees and some elms and cedars planted by the poet himself. A space once laid out in winding walks and beautiful shrubberies is now a potatoe field! The present proprietor, Mr. Young, is a wholesale tea-dealer. Even the bones of the poet, it is said, have been disturbed. The skull of Pope, according to William Howitt, is now in the private collection of a phrenologist! The manner in which it was obtained, he says, is this:--On some occasion of alteration in the church at Twickenham, or burial of some one in the same spot, the coffin of Pope was disinterred, and opened to see the state of the remains. By a bribe of £50 to the Sexton, possession of the skull was obtained for one night; another skull was then returned instead of the poet's.

It has been stated that the French term *Ferme Ornée* was first used in England by Shenstone. It exactly expressed the character of his grounds. Mr. Repton said that he never strolled over the scenery of the Leasowes without lamenting the constant disappointment to which Shenstone exposed himself by a vain attempt to unite the incompatible objects of ornament and profit. "Thus," continued Mr. Repton, "the poet lived under the continual mortification of disappointed hope, and with a mind exquisitely sensible, he felt equally the sneer of the great man at the magnificence of his attempt and the ridicule of the farmer at the misapplication of his paternal acres." The "sneer of the great man." is perhaps an allusion to what Dr. Johnson says of Lord Lyttelton:--that he "looked with disdain" on "the petty State" of his neighbour. "For a while," says Dr. Johnson, "the inhabitants of Hagley affected to tell their acquaintance of the little fellow that was trying to make himself admired; but when by degrees the Leasowes forced themselves into notice, they took care to defeat the curiosity which they could not suppress, by conducting their visitants perversely to inconvenient points of view, and introducing them at the wrong end of a walk to detect a deception; injuries of which Shenstone would heavily complain." Mr. Graves, the zealous friend of Shenstone, indignantly denies that any of the Lyttelton family had evinced so ungenerous a feeling towards the proprietor of the Leasowes who though his "empire" was less "spacious and opulent" had probably a larger share of true taste than even the proprietor of Hagley, the Lyttelton domain--though Hagley has been much, and I doubt not, deservedly, admired.[023]

Dr. Johnson states that Shenstone's expenses were beyond his means,-- that he spent his estate in adorning it--that at last the clamours of creditors "overpowered the lamb's bleat and the linnet's song; and that his groves were haunted by beings very different from fauns and fairies." But this is gross exaggeration. Shenstone was occasionally, indeed, in slight pecuniary difficulties, but he could always have protected himself from the intrusion of the myrmidons of the law by raising money on his estate; for it appears that after the payment of all his debts, he left legacies to his friends and annuities to his servants.

Johnson himself is the most scornful of the critics upon Shenstone's rural pursuits. "The pleasure of Shenstone," says the Doctor, "was all in his eye: he valued what he valued merely for its looks. Nothing raised his indignation more than to ask if there were any fishes in his water." Dr. Johnson would have seen no use in the loveliest piece of running water in the world if it had contained nothing that he could masticate! Mrs. Piozzi says of him, "The truth is, he hated to hear about prospects and views, and laying out grounds and taste in gardening." "That was the best garden," he said, "which produced most roots and fruits; and that water was most to be prized which contained most fish." On this principle of the valuelessness of those pleasures which enter the mind through the eye, Dr. Johnson should have blamed the lovers of painting for dwelling with such fond admiration on the canvas of his friend Sir Joshua Reynolds. In point of fact, Dr. Johnson had no more sympathy with the genius of the painter or the musician than with that of the Landscape gardener, for he had neither an eye nor an ear for Art. He wondered how any man could be such a fool as to be moved to tears by music, and observed, that, "one could not fill one's belly with hearing soft murmurs or looking at rough cascades." No; the loveliness of nature does not satisfy the thirst and hunger of the body, but it *does* satisfy the thirst and hunger of the soul. No one can find wheaten bread or wine or venison or beef or plum-pudding or turtle-soup in mere sounds and sights, however exquisite--neither can any one find such substantial diet within the boards of a book-- no not even on the pages of Shakespeare, or even those of the Bible itself,--but men can find in sweet music and lovely scenery and good books something infinitely more precious than all the wine, venison, beef, or plum- pudding, or turtle-soup that could be swallowed during a long life by the most craving and capacious alderman of London! Man is of a dual nature: he is not all body. He has other and far higher wants and enjoyments than the purely physical--and these nobler appetites are gratified by the charms of nature and the creations of inspired genius.

Dr. Johnson's gastronomic allusions to nature recal the old story of a poet pointing out to a utilitarian friend some white lambs frolicking in a meadow. "Aye," said, the other, "only think of a quarter of one of them with asparagus and mint sauce!" The story is by some supposed to have had a Scottish origin, and a prosaic North Briton is made to say that the pretty little lambs, sporting amidst the daisies and buttercups, would "*mak braw pies*."

A profound feeling for the beautiful is generally held to be an essential quality in the poet. It is a curious fact, however, that there are some who aspire to the rank of poet, and have their claims allowed, who yet cannot be said to be poetical in their nature--for how can that nature be, strictly speaking, *poetical* which denies the sentiment of Keats, that

> A thing of beauty is a joy for ever?

Both Scott and Byron very earnestly admired Dr. Johnson's "*London*" and "*The Vanity of Human Wishes*." Yet the sentiments just quoted from the author of those productions are far more characteristic of a utilitarian philosopher than of one who has been endowed by nature with

> The vision and the faculty divine,

and made capable, like some mysterious enchanter, of

> Clothing the palpable and the familiar With golden exhalations of the dawn.

Crabbe, also a prime favorite with the authors of the *Lay of the Last Minstrel*, and *Childe Harold*, is recorded by his biographer--his own son--to have exhibited "a remarkable indifference to all the proper objects of taste;" to have had "no real love for painting, or music, or architecture or for what a painter's eye considers as the beauties of landscape." "In botany, grasses, the most *useful* but the least ornamental, were his favorites." "He never seemed to be captivated with the mere beauty of natural objects or even to catch any taste for the arrangement of his specimens. Within, the house was a kind of scientific confusion; in the garden the usual showy foreigners gave place to the most scarce flowers, especially to the rarer weeds, of Britain; and were scattered here and there only for preservation. In fact he neither loved order for its own sake nor had any very high opinion of that passion in others."[024] Lord Byron described Crabbe to be

> Though nature's sternest painter, yet *the best*.

What! was he a better painter of nature than Shakespeare? The truth is that Byron was a wretched critic, though a powerful poet. His praises and his censures were alike unmeasured.

> His generous ardor no cold medium knew.

He seemed to recognize no great general principles of criticism, but to found all his judgments on mere prejudice and passion. He thought Cowper "no poet," pronounced Spenser "a dull fellow," and placed Pope above Shakespeare. Byron's line on Crabbe is inscribed on the poet's tombstone at Trowbridge. Perhaps some foreign visitor on reading the inscription may be surprised at his own ignorance when he learns that it is not the author of *Macbeth* and *Othello* that he is to regard as the best painter of nature that England has produced, but the author of the *Parish Register* and the *Tales of the Hall*. Absurd and indiscriminate laudations of this kind confound all intellectual distinctions and make criticism ridiculous. Crabbe is unquestionably a vigorous and truthful writer, but he is not the *best* we have, in any sense of the word.

Though Dr. Johnson speaks so contemptuously of Shenstone's rural pursuits, he could not help acknowledging that when the poet began "to point his prospects, to diversify his surface, to entangle his walks and to wind his waters," he did all this with such judgment and fancy as "made his little domain the envy of the great and the admiration of the skilful; a place to be visited by travellers, and copied by designers."

Mason, in his *English Garden*, a poem once greatly admired, but now rarely read, and never perhaps with much delight, does justice to the taste of the Poet of the Leasowes.

> Nor, Shenstone, thou Shalt pass without thy meed, thou son of peace! Who knew'st, perchance, to harmonize thy shades Still softer than thy song; yet was that song Nor rude nor inharmonious when attuned To pastoral plaint, or tale of slighted love.

English pleasure-gardens have been much imitated by the French. Viscomte Girardin, at his estate of Ermenonville, dedicated an inscription in amusing French-English to the proprietor of the Leasowes--

> THIS PLAIN STONE TO WILLIAM SHENSTONE; IN HIS WRITINGS HE DISPLAYED A MIND NATURAL; AT LEASOWES HE LAID ARCADIAN GREENS RURAL.

The Viscomte, though his English composition was so quaint and imperfect, was an elegant writer in his own language, and showed great taste and skill in laying out his grounds. He had visited England, and carefully studied our modern style of gardening. He had personally consulted Shenstone, Mason, Whateley and other English authors on subjects of rural taste. He published an eloquent description of his own estate. His famous friend Rousseau wrote the preface to it. The book was translated into English. Rousseau spent his last days at Ermenonville and was buried there in what is called *The Isle of Poplars*. The garden is now in a neglected state, but the tomb of Rousseau remains uninjured, and is frequently visited by the admirers of his genius.

"Dr. Warton," says Bowles, "mentions Milton and Pope as the poets to whom English Landscape is indebted, but *he forgot poor Shenstone*." A later writer, however, whose sympathy for genius communicates such a charm to all his anecdotes and comments in illustration of the literary character, has devoted a chapter of his *Curiosities of Literature* to a notice of the rural tastes of the proprietor of the Leasowes. I must give a brief extract from it.

"When we consider that Shenstone, in developing his fine pastoral ideas in the Leasowes, educated the nation into that taste for landscape- gardening, which has become the model of all Europe, this itself constitutes a claim on the gratitude of posterity. Thus the private pleasures of a man of genius may become at length those of a whole people. The creator of this new taste appears to have received far less notice than he merited. The name of Shenstone does not appear in the Essay on Gardening, by Lord Orford; even the supercilious Gray only bestowed a ludicrous image on these pastoral scenes, which, however, his friend Mason has celebrated; and the genius of Johnson, incapacitated by nature to touch on objects of rural fancy, after describing some of the offices of the landscape designer, adds, that 'he will not inquire whether they demand any great powers of mind.' Johnson, however, conveys to us his own feelings, when he immediately expresses them under the character of 'a sullen and surly speculator.' The anxious life of Shenstone would indeed have been remunerated, could he have read the enchanting eulogium of Whateley on the Leasowes; which, said he, 'is a perfect picture of his mind--simple, elegant and amiable; and will always suggest a doubt whether the spot inspired his verse, or whether in the scenes which he formed, he only realised the pastoral images which abound in his songs.' Yes! Shenstone had been delighted could he have heard that Montesquieu, on his return home, adorned his 'Chateau Gothique, mais orné de bois charmans, don't j'ai pris l'idée en Angleterre;' and Shenstone, even with his modest and timid nature, had been proud to have witnessed a noble foreigner, amidst memorials dedicated to Theocritus and Virgil, to Thomson and Gesner, raising in his grounds an inscription, in bad English, but in pure taste, to Shenstone himself; for having displayed in his writings 'a mind natural,' and in his Leasowes 'laid Arcadian greens rural;' and recently Pindemonte has traced the taste of English gardening to Shenstone. A man of genius sometimes receives from foreigners, who are placed out of the prejudices of his compatriots, the tribute of posterity!"

"The Leasowes," says William Howitt, "now belongs to the Atwood family; and a Miss Atwood resides there occasionally. But the whole place bears the impress of desertion and neglect. The house has a dull look; the same heavy spirit broods over the lawns and glades: And it is only when you survey it from a distance, as when approaching Hales-Owen from Hagley, that the whole presents an aspect of unusual beauty."

Shenstone was at least as proud of his estate of the Leasowes as was Pope of his Twickenham Villa--perhaps more so. By mere men of the world, this pride in a garden may be regarded as a weakness, but if it be a weakness it is at least an innocent and inoffensive one, and it has been associated with the noblest intellectual endowments. Pitt and Fox and Burke and Warren Hastings were not weak men, and yet were they all extremely proud of their gardens. Every one, indeed, who takes an active interest in the culture and embellishment of his garden, finds his pride in it and his love for it increase daily. He is delighted to see it flourish and improve beneath his care. Even the humble mechanic, in his fondness for a garden, often indicates a feeling for the beautiful, and a genial nature. If a rich man were openly to boast of his plate or his equipages, or a literary man of his essays or his sonnets, as lovers of flowers boast of their geraniums or dahlias or rhododendrons, they would disgust the most indulgent hearer. But no one is shocked at the exultation of a gardener, amateur or professional, when in the fulness of his heart he descants upon the unrivalled beauty of his favorite flowers:

'Plants of his hand, and children of his care.'

"I have made myself two gardens," says Petrarch, "and I do not imagine that they are to be equalled in all the world. I should feel myself inclined to be angry with fortune if there were any so beautiful out of Italy." "I wish," says poor Kirke White writing to a friend, "I wish you to have a taste of these (rural) pleasures with me, and if ever I should live to be blessed with a quiet parsonage, and *another great object of my ambition--a garden*, I have no doubt but we shall be for some short intervals at least two quite contented bodies." The poet Young, in the latter part of his life, after years of vain hopes and worldly struggles, gave himself up almost entirely to the sweet seclusion of a garden; and that peace and repose which cannot be found in courts and political cabinets, he found at last

> In sunny garden bowers Where vernal winds each tree's low tones awaken, And buds and bells with changes mark the hours.

He discovered that it was more profitable to solicit nature than to flatter the great.

> For Nature never did betray The heart that loved her.

People of a poetical temperament--all true lovers of nature--can afford, far better than more essentially worldly beings, to exclaim with Thomson.

> I care not Fortune what you me deny, You cannot bar me of free Nature's grace, You cannot shut the windows of the sky Through which Aurora shows her brightening face: You cannot bar my constant feet to trace The woods and lawns and living streams at eve: Let health my nerves and finer fibres brace, And I their toys to the *great children* leave:-- Of fancy, reason, virtue, nought can me bereave.

The pride in a garden laid out under one's own directions and partly cultivated by one's own hand has been alluded to as in some degree unworthy of the dignity of manhood, not only by mere men of the world, or silly coxcombs, but by people who should have known better. Even Sir William Temple, though so enthusiastic about his fruit-trees, tells us that he will not enter upon any account of *flowers*, having only pleased himself with seeing or smelling them, and not troubled himself with the care of them, which he observes "*is more the ladies part than the men's.*" Sir William makes some amends for this almost contemptuous allusion to flowers in particular by his ardent appreciation of the use of gardens and gardening in general. He thus speaks of their attractions and advantages: "The sweetness of the air, the pleasantness of the smell, the verdure of plants, the cleanness and lightness of food, the exercise of working or walking, but above all, the exemption from cares and solicitude, seem equally to favor and improve both contemplation and health, the enjoyment of sense and imagination, and thereby the quiet and ease of the body and mind." Again: "As gardening has been the inclination of kings and the choice of philosophers, so it has been the common favorite of public and private men, a pleasure of the greatest and the care of the meanest; and indeed *an employment and a possession for which no man is too high or too low.*" This is just and liberal; though I can hardly help still feeling a little sore at Sir William's having implied in the passage previously quoted, that the care of flowers is but a feminine occupation. As an elegant amusement, it is surely equally well fitted for all lovers of the beautiful, without reference to their sex.

It is not women and children only who delight in flower-gardens. Lord Bacon and William Pitt and the Earl of Chatham and Fox and Burke and Warren Hastings--all lovers of flowers--were assuredly not men of frivolous minds or of feminine habits. They were always eager to exhibit to visitors the beauty of their parterres. In his declining years the stately John Kemble left the stage for his garden. That sturdy English yeoman, William Cobbett, was almost as proud of his beds of flowers as of the pages of his *Political Register*. He thus speaks of gardening:

"Gardening is a source of much greater profit than is generally imagined; but, merely as an amusement or recreation it is a thing of very great value. It is not only compatible with but favorable to the study of any art or science; it is conducive to health by means of the irresistible temptation which it offers to early rising; to the stirring abroad upon one's legs, for a man may really ride till he cannot walk, sit till he cannot stand, and lie abed till he cannot get up. It tends to turn the minds of youth from amusements and attachments of a frivolous and vicious nature, it is a taste which is indulged at home; it tends to make home pleasant, and to endear to us the spot on which it is our lot to live,--and as to the *expenses* attending it, what are all these expenses compared with those of the short, the unsatisfactory, the injurious enjoyment of the card-table, and the rest of those amusements which are sought from the town." *Cobbett's English Gardener.*

"Other fine arts," observes Lord Kames, "may be perverted to excite irregular and even vicious emotions: but gardening, which inspires the purest and most refined pleasures, cannot fail to promote every good affection. The gaiety and harmony of mind it produceth, inclining the spectator to communicate his satisfaction to others, and to make them happy as he is himself, tend naturally to establish in him a habit of humanity and benevolence."

Every thoughtful mind knows how much the face of nature has to do with human happiness. In the open air and in the midst of summer-flowers, we often feel the truth of the observation that "a fair day is a kind of sensual pleasure, and of all others the most innocent." But it is also something more, and better. It kindles a spiritual delight. At such a time and in such a scene every observer capable of a religious emotion is ready to exclaim--

> Oh! there is joy and happiness in every thing I see, Which bids my soul rise up and bless the God that blesses me

Anon.

The amiable and pious Doctor Carey of Serampore, in whose grounds sprang up that dear little English daisy so beautifully addressed by his poetical proxy, James Montgomery of Sheffield, in the stanzas commencing:--

> Thrice welcome, little English flower! My mother country's
> white and red--

was so much attached to his Indian garden, that it was always in his heart in the intervals of more important cares. It is said that he remembered it even upon his death-bed, and that it was amongst his last injunctions to his friends that they should see to its being kept up with care. He was particularly anxious that the hedges or railings should always be in such good order as to protect his favorite shrubs and flowers from the intrusion of Bengalee cattle.

A garden is a more interesting possession than a gallery of pictures or a cabinet of curiosities. Its glories are never stationary or stale. It has infinite variety. It is not the same to-day as it was yesterday. It is always changing the character of its charms and always increasing them in number. It delights all the senses. Its pleasures are not of an unsocial character; for every visitor, high or low, learned or illiterate, may be fascinated with the fragrance and beauty of a garden. But shells and minerals and other curiosities are for the man of science and the connoisseur. And a single inspection of them is generally sufficient: they never change their aspect. The Picture-Gallery may charm an instructed eye but the multitude have little relish for human Art, because they rarely understand it:--while the skill of the Great Limner of Nature is visible in every flower of the garden even to the humblest swain.

It is pleasant to read how the wits and beauties of the time of Queen Anne used to meet together in delightful garden-retreats, 'like the companies in Boccaccio's Decameron or in one of Watteau's pictures.' Ritchings Lodge, for instance, the seat of Lord Bathurst, was visited by most of the celebrities of England, and frequently exhibited bright groups of the polite and accomplished of both sexes; of men distinguished for their heroism or their genius, and of women eminent for their easy and elegant conversation, or for gaiety and grace of manner, or perfect loveliness of face and form-- all in harmonious union with the charms of nature. The gardens at Ritchings were enriched with Inscriptions from the pens of Congreve and Pope and Gay and Addison and Prior. When the estate passed into the possession of the Earl of Hertford, his literary lady devoted it to the Muses. "She invited every summer," says Dr. Johnson, "some poet into the country to hear her verses and assist her studies." Thomson, who praises her so lavishly in his "Spring," offended her ladyship by allowing her too clearly to perceive that he was resolved not to place himself in the dilemma of which Pope speaks so feelingly with reference to other poetasters.

> Seized and tied down to judge, how wretched I, Who can't be silent, and who will not lie. I sit with sad civility, I read With honest anguish and an aching head.

But though "the bard more fat than bard beseems" was restive under her ladyship's "poetical operations," and too plainly exhibited a desire to escape the infliction, preferring the Earl's claret to the lady's rhymes, she should have been a little more generously forgiving towards one who had already made her immortal. It is stated, that she never repeated her invitation to the Poet of the Seasons, who though so impatient of the sound of her tongue when it "rolled" her own "raptures," seems to have been charmed with her *at a distance*--while meditating upon her excellencies in the seclusion of his own study. The compliment to the Countess is rather awkwardly wedged in between descriptions of "gentle Spring" with her "shadowing roses" and "surly Winter" with his "ruffian blasts." It should have commenced the poem.

> O Hertford, fitted or to shine in courts With unaffected grace, or walk the plain, With innocence and meditation joined In soft assemblage, listen to my song, Which thy own season paints; when nature all Is blooming and benevolent like thee.

Thomson had no objection to strike off a brief compliment in verse, but he was too indolent to keep up *in propriâ personâ* an incessant fire of compliments, like the *bon bons* at a Carnival. It was easier to write her praises than listen to her verses. Shenstone seems to have been more pliable. He was personally obsequious, lent her recitations an attentive ear, and was ever ready with the expected commendation. It is not likely that her ladyship found much, difficulty in collecting around her a crowd of critics more docile than Thomson and quite as complaisant as Shenstone. Let but a *Countess*

Once own the happy lines, How the wit brightens, how the style refines!

Though Thomson's first want on his arrival in London from the North was a pair of shoes, and he lived for a time in great indigence, he was comfortable enough at last. Lord Lyttleton introduced him to the Prince of Wales (who professed himself the patron of literature) and when his Highness questioned him about the state of his affairs, Thomson assured him that they "were in a more poetical posture than formerly." The prince bestowed upon the poet a pension of a hundred pounds a year, and when his friend Lord Lyttleton was in power his Lordship obtained for him the office of Surveyor General of the Leeward Islands. He sent a deputy there who was more trustworthy than Thomas Moore's at Bermuda. Thomson's deputy after deducting his own salary remitted his principal three hundred pounds per annum, so that the bard 'more fat than bard beseems' was not in a condition to grow thinner, and could afford to make his cottage a Castle of Indolence. Leigh Hunt has versified an anecdote illustrative of Thomson's luxurious idleness. He who could describe "*Indolence*" so well, and so often appeared in the part himself,

> Slippered, and with hands, Each in a waistcoat pocket, (so that all Might yet repose that could) was seen one morn Eating a wondering peach from off the tree.

A little summer-house at Richmond which Thomson made his study is still preserved, and even some articles of furniture, just as he left them.[025] Over the entrance is erected a tablet on which is the following inscription:

> HERE THOMSON SANG THE SEASONS AND THEIR CHANGE.

Thomson was buried in Richmond Church. Collins's lines to his memory, beginning

> In yonder grave a Druid lies,

are familiar to all readers of English poetry.

Richmond Hill has always been the delight not of poets only but of painters. Sir Joshua Reynolds built a house there, and one of the only three landscapes which seem to have survived him, is a view from the window of his drawing-room. Gainsborough was also a resident in Richmond. Richmond gardens laid out or rather altered by Brown, are now united with those of Kew.

Savage resided for some time at Richmond. It was the favorite haunt of Collins, one of the most poetical of poets, who, as Dr. Johnson says, "delighted to rove through the meanders of enchantment, to gaze on the magnificence of golden palaces, to repose by the waterfalls of Elysian gardens." Wordsworth composed a poem upon the Thames near Richmond in remembrance of Collins. Here is a stanza of it.

> Glide gently, thus for ever glide, O Thames, that other bards may see As lovely visions by thy side As now fair river! come to me; O glide, fair stream for ever so, Thy quiet soul on all bestowing, Till all our minds for ever flow As thy deep waters now are flowing.

Thomson's description of the scenery of Richmond Hill perhaps hardly does it justice, but the lines are too interesting to be omitted.

> Say, shall we wind Along the streams? or walk the smiling mead? Or court the forest-glades? or wander wild Among the waving harvests? or ascend, While radiant Summer opens all its pride, Thy hill, delightful Shene[026]? Here let us sweep The boundless landscape now the raptur'd eye, Exulting swift, to huge Augusta send, Now to the sister hills[027] that skirt her plain, To lofty Harrow now, and now to where Majestic Windsor lifts his princely brow In lovely contrast to this glorious view Calmly magnificent, then will we turn To where the silver Thames first rural grows There let the feasted eye unwearied stray, Luxurious, there, rove through the pendent woods That nodding hang o'er Harrington's retreat, And stooping thence to Ham's embowering walks, Beneath whose shades, in spotless peace retir'd, With her the

pleasing partner of his heart, The worthy Queensbury yet laments his Gay, And polish'd Cornbury woos the willing Muse Slow let us trace the matchless vale of Thames Fair winding up to where the Muses haunt In Twit nam's bowers, and for their Pope implore The healing god[028], to loyal Hampton's pile, To Clermont's terrass'd height, and Esher's groves; Where in the sweetest solitude, embrac'd By the soft windings of the silent Mole, From courts and senates Pelham finds repose Enchanting vale! beyond whate'er the Muse Has of Achaia or Hesperia sung! O vale of bliss! O softly swelling hills! On which the *Power of Cultivation* lies, And joys to see the wonders of his toil.

The Revd. Thomas Maurice wrote a poem entitled *Richmond Hill*, but it contains nothing deserving of quotation after the above passage from Thomson. In the *English Bards and Scotch Reviewers* the labors of Maurice are compared to those of Sisyphus

So up thy hill, ambrosial Richmond, heaves Dull Maurice, all his granite weight of leaves.

Towards the latter part of the last century the Empress of Russia (Catherine the Second) expressed in a French letter to Voltaire her admiration of the style of English Gardening.[029] "I love to distraction," she writes, "the present English taste in gardening. Their curved lines, their gentle slopes, their pieces of water in the shape of lakes, their picturesque little islands. I have a great contempt for straight lines and parallel walks. I hate those fountains which torture water into forms unknown to nature. I have banished all the statues to the vestibules and to the galleries. In a word English taste predominates in my *plantomanie*."[030]

I omitted when alluding to those Englishmen in past times who anticipated the taste of the present day in respect to laying out grounds, to mention the ever respected name of John Evelyn, and as all other writers before me, I believe, who have treated upon gardening, have been guilty of the same oversight, I eagerly make his memory some slight amends by quoting the following passage from one of his letters to his friend Sir Thomas Browne.

"I might likewise hope to refine upon some particulars, especially concerning the ornaments of gardens, which I shall endeavor so to handle as that they may become useful and practicable, as well as magnificent, and that persons of all conditions and faculties, which delight in gardens, may therein encounter something for their owne advantage. The modell, which I perceive you have seene, will aboundantly testifie my abhorrency of those painted and formal projections of our cockney gardens and plotts, which appeare like gardens of past-board and marchpane, and smell more of paynt then of flowers and verdure; our drift is a noble, princely, and universal Elysium, capable of all the amoenities that can naturally be introduced into gardens of pleasure, and such as may stand in competition with all the august designes and stories of this nature, either of antient or moderne tymes; yet so as to become useful and significant to the least pretences and faculties. We will endeavour to shew how the air and genious of gardens operat upon humane spirits towards virtue and sanctitie: I mean in a remote, preparatory and instrumentall working. How caves, grotts, mounts, and irregular ornaments of gardens do contribute to contemplative and philosophicall enthusiasme; how *elysium, antrum, nemus, paradysus, hortus, lucus,* &c., signifie all of them *rem sacram it divinam*; for these expedients do influence the soule and spirits of men, and prepare them for converse with good angells; besides which, they contribute to the lesse abstracted pleasures, phylosophy naturall; and longevitie: and I would have not onely the elogies and effigie of the antient and famous garden heroes, but a society of the *paradisi cultores* persons of antient simplicity, Paradisean and Hortulan saints, to be a society of learned and ingenuous men, such as Dr. Browne, by whome we might hope to redeeme the tyme that has bin lost, in pursuing *Vulgar Errours*, and still propagating them, as so many bold men do yet presume to do."

The English style of landscape-gardening being founded on natural principles must be recognized by true taste in all countries. Even in Rome, when art was most allowed to predominate over nature, there were occasional instances of that correct feeling for rural beauty which the English during the last century and a half have exhibited more conspicuously than other nations. Atticus preferred Tully's villa at Arpinum to all his other villas; because at Arpinum,

Nature predominated over art. Our Kents and Browns[031] never expressed a greater contempt, than was expressed by Atticus, for all formal and artificial decorations of natural scenery.

The spot where Cicero's villa stood, was, in the time of Middleton, possessed by a convent of monks and was called the Villa of St. Dominic. It was built, observes Mr. Dunlop, in the year 1030, from the fragments of the Arpine Villa!

> Art, glory, Freedom, fail--but Nature still is fair.

"Nothing," says Mr. Kelsall, "can be imagined finer than the surrounding landscape. The deep azure of the sky, unvaried by a single cloud--Sora on a rock at the foot of the precipitous Appennines--both banks of the Garigliano covered with vineyards--the *fragor aquarum*, alluded to by Atticus in his work *De Legibus*--the coolness, the rapidity and ultramarine hue of the Fibrenus--the noise of its cataracts--the rich turquoise color of the Liris--the minor Appennines round Arpino, crowned with umbrageous oaks to the very summits--present scenery hardly elsewhere to be equalled, certainly not to be surpassed, even in Italy."

This description of an Italian landscape can hardly fail to charm the imagination of the coldest reader; but after all, I cannot help confessing to so inveterate a partiality for dear old England as to be delighted with the compliment which Gray, the poet, pays to English scenery when he prefers it to the scenery of Italy. "Mr. Walpole," writes the poet from Italy, "says, our *memory* sees more than our eyes in this country. This is extremely true, since for *realities* WINDSOR or RICHMOND HILL is infinitely preferable to ALBANO or FRESCATI."

Sir Walter Scott, with all his patriotic love for his own romantic land, could not withhold his tribute to the loveliness of Richmond Hill,--its "*unrivalled landscape*" its "*sea of verdure.*"

> "They" (The Duke of Argyle and Jeanie Deans) "paused for a moment on the brow of a hill, to gaze on the unrivalled landscape it presented. A huge sea of verdure, with crossing and intersecting promontories of massive and tufted groves was

tenanted by numberless flocks and herds which seemed to wander unrestrained and unbounded through the rich pastures. The Thames, here turreted with villas, and there garlanded with forests, moved on slowly and placidly, like the mighty monarch of the scene, to whom all its other beauties were but accessaries, and bore on its bosom an hundred barks and skiffs whose white sails and gaily fluttering pennons gave life to the whole." *The Heart of Mid-Lothian.*

It must of course be admitted that there are grander, more sublime, more varied and extensive prospects in other countries, but it would be difficult to persuade me that the richness of English verdure could be surpassed or even equalled, or that any part of the world can exhibit landscapes more truly *lovely* and *loveable*, than those of England, or more calculated to leave a deep and enduring impression upon the heart. Mr. Kelsall speaks of an Italian sky "*uncovered by a single cloud*," but every painter and poet knows how much variety and beauty of effect are bestowed upon hill and plain and grove and river by passing clouds; and even our over-hanging vapours remind us of the veil upon the cheek of beauty; and ever as the sun uplifts the darkness the glory of the landscape seems renewed and freshened. It would cheer the saddest heart and send the blood dancing through the veins, to behold after a dull misty dawn, the sun break out over Richmond Hill, and with one broad light make the whole landscape smile; but I have been still more interested in the prospect when on a cloudy day the whole "sea of verdure" has been swayed to and fro into fresher life by the fitful breeze, while the lights and shadows amidst the foliage and on the lawns have been almost momentarily varied by the varying sky. These changes fascinate the eye, keep the soul awake, and save the scenery from the comparatively monotonous character of landscapes in less varying climes. And for my own part, I cordially echo the sentiment of Wordsworth, who when conversing with Mrs. Hemans about the scenery of the Lakes in the North of England, observed: "I would not give up the mists that *spiritualize* our mountains for all the blue skies of Italy."

Though Mrs. Stowe, the American authoress already quoted as one of the admirers of England, duly appreciates the natural grandeur of her own land, she was struck with admiration and delight at the aspect of our English landscapes. Our trees, she observes, "are of an order of nobility and they wear their crowns right kingly." "Leaving out of account," she adds, "our *mammoth arboria*, the English Parks have trees as fine and effective as ours, and when I say their trees are of an order of nobility, I mean that they (the English) pay a reverence to them such as their magnificence deserves."

Walter Savage Landor, one of the most accomplished and most highly endowed both by nature and by fortune of our living men of letters, has done, or rather has tried to do, almost as much for his country in the way of enriching its collection of noble trees as Evelyn himself. He laid out £70,000 on the improvement of an estate in Monmouthshire, where he planted and fenced half a million of trees, and had a million more ready to plant, when the conduct of some of his tenants, who spitefully uprooted them and destroyed the whole plantation, so disgusted him with the place, that he razed to the ground the house which had cost him £8,000, and left the country. He then purchased a beautiful estate in Italy, which is still in possession of his family. He himself has long since returned to his native land. Landor loves Italy, but he loves England better. In one of his *Imaginary Conversations* he tells an Italian nobleman:

"The English are more zealous of introducing new fruits, shrubs and plants, than other nations; you Italians are less so than any civilized one. Better fruit is eaten in Scotland than in the most fertile and cultivated parts of your peninsula. *As for flowers, there is a greater variety in the worst of our fields than in the best of your gardens.* As for shrubs, I have rarely seen a lilac, a laburnum, a mezereon, in any of them, and yet they flourish before almost every cottage in our poorest villages."

"We wonder in England, when we hear it related by travellers, that peaches in Italy are left under the trees for swine; but, when we ourselves come into the country, our wonder is rather that the swine do not leave them for animals less nice."

Landor acknowledges that he has eaten better pears and cherries in Italy than in England, but that all the other kinds of fruitage in Italy appeared to him unfit for dessert.

The most celebrated of the private estates of the present day in England is Chatsworth, the seat of the Duke of Devonshire. The mansion, called the Palace of the Peak, is considered one of the most splendid residences in the land. The grounds are truly beautiful and most carefully attended to. The elaborate waterworks are perhaps not in the severest taste. Some of them are but costly puerilities. There is a water-work in the form of a tree that sends a shower from every branch on the unwary visitor, and there are snakes that spit forth jets upon him as he retires. This is silly trifling: but ill adapted to interest those who have passed their teens; and not at all an agreeable sort of hospitality in a climate like that of England. It is in the style of the water-works at Versailles, where wooden soldiers shoot from their muskets vollies of water at the spectators.[032]

It was an old English custom on certain occasions to sprinkle water over the company at a grand entertainment. Bacon, in his Essay on Masques, seems to object to getting drenched, when he observes that "some sweet odours suddenly coming forth, *without any drops falling*, are in such a company as there is steam and heat, things of great pleasure and refreshment." It was a custom also of the ancient Greeks and Romans to sprinkle their guests with fragrant waters. The Gascons had once the same taste: "At times," says Montaigne, "from the bottom of the stage, they caused sweet-scented waters to spout upwards and dart their thread to such a prodigious height, as to sprinkle and perfume the vast multitudes of spectators." The Native gentry of India always slightly sprinkle their visitors with rose-water. It is flung from a small silver utensil tapering off into a sort of upright spout with a pierced top in the fashion of that part of a watering pot which English gardeners call the *rose*.

The finest of the water-works at Chatsworth is one called the *Emperor Fountain* which throws up a jet 267 feet high. This height exceeds that of any fountain in Europe. There is a vast Conservatory on the estate, built of glass by Sir Joseph Paxton, who designed and constructed the Crystal Palace. His experience in the building of conservatories no doubt suggested to him the idea of the splendid glass edifice in Hyde Park. The conservatory at Chatsworth required 70,000 square feet of glass. Four miles of iron tubing are used in heating the building. There is a broad carriage way running right through the centre of the conservatory.[033] This conservatory is peculiarly rich in exotic plants of all kinds, collected at an enormous cost. This most princely estate, contrasted with the little cottages and cottage-gardens in the neighbourhood, suggested to Wordsworth the following sonnet.

CHATSWORTH.

> Chatsworth! thy stately mansion, and the pride Of thy domain, strange contrast do present To house and home in many a craggy tent Of the wild Peak, where new born waters glide Through fields whose thrifty occupants abide As in a dear and chosen banishment With every semblance of entire content; So kind is simple Nature, fairly tried! Yet he whose heart in childhood gave his troth To pastoral dales, then set with modest farms, May learn, if judgment strengthen with his growth, That not for Fancy only, pomp hath charms; And, strenuous to protect from lawless harms The extremes of favored life, may honour both.

The two noblest of modern public gardens in England are those at Kensington and Kew. Kensington Gardens were begun by King William the III, but were originally only twenty-six acres in extent. Queen Anne added thirty acres more. The grounds were laid out by the well-known garden-designers, London and Wise.[034] Queen Caroline, who formed the Serpentine River by connecting several detached pieces of water into one, and set the example of a picturesque deviation from the straight line,[035] added from Hyde Park no less than three hundred acres which were laid out by Bridgeman. This was a great boon to the Londoners. Horace Walpole says that Queen Caroline at first proposed to shut up St. James's Park and convert it into a private garden for herself, but when she asked Sir Robert Walpole what it would cost, he answered--"Only three Crowns." This changed her intentions.

The reader of Pope will remember an allusion to the famous Ring in Hyde Park. The fair Belinda was sometimes attended there by her guardian Sylphs:

> The light militia of the lower sky.

They guarded her from 'the white-gloved beaux,'

> These though unseen are ever on the wing, Hang o'er the box, *and hover o'er the Ring.*

It was here that the gallantries of the "Merry Monarch" were but too often exhibited to his people. "After dinner," says the right garrulous Pepys in his journal, "to Hyde Parke; at the Parke was the King, and in another Coach, Lady Castlemaine, they greeting one another at every turn."

The Gardens at Kew "Imperial Kew," as Darwin styles it, are the richest in the world. They consist of one hundred and seventy acres. They were once private gardens, and were long in the possession of Royalty, until the accession of Queen Victoria, who opened the gardens to the public and placed them under the control of the Commissioners of Her Majesty's Woods and Forests, "with a view of rendering them available to the general good."

> She hath left you all her walks, Her private arbors and new planted orchards On this side Tiber. She hath left them you And to your heirs for ever; common pleasures To walk abroad and recreate yourselves.

They contain a large Palm-house built in 1848.[036] The extent of glass for covering the building is said to be 360,000 square feet. My Mahomedan readers in Hindostan, (I hope they will be numerous,) will perhaps be pleased to hear that there is an ornamental mosque in these gardens. On each of the doors of this mosque is an Arabic inscription in golden characters, taken from the Koran. The Arabic has been thus translated:--

> LET THERE BE NO FORCE IN RELIGION. THERE IS NO OTHER GOD EXCEPT THE DEITY. MAKE NOT ANY LIKENESS UNTO GOD.

The first sentence of the translation is rather ambiguously worded. The sentiment has even an impious air: an apparent meaning very different from that which was intended. Of course the original text *means*, though the English translator has not expressed that meaning--"Let there be no force *used* in religion."

When William Cobbett was a boy of eleven years of age he worked in the garden of the Bishop of Winchester at Farnham. Having heard much of Kew gardens he resolved to change his locality and his master. He started off for Kew, a distance of about thirty miles, with only thirteen pence in his pocket. The head gardener at Kew at once engaged his services. A few days after, George the Fourth, then Prince of Wales, saw the boy sweeping the lawns, and laughed heartily at his blue smock frock and long red knotted garters. But the poor gardener's boy became a public writer, whose productions were not exactly calculated to excite the merriment of princes.

Most poets have a painter's eye for the disposition of forms and colours. Kent's practice as a painter no doubt helped to make him what he was as a landscape-gardener. When an architect was consulted about laying out the grounds at Blenheim he replied, "you must send for a landscape-painter:" he might have added--"*or a poet.*"

Our late Laureate, William Wordsworth, exhibited great taste in his small garden at Rydal Mount. He said of himself--very truly though not very modestly perhaps,--but modesty was never Wordsworth's weakness-- that nature seemed to have fitted him for three callings--that of the poet, the critic on works of art, and the landscape-gardener. The poet's nest--(Mrs. Hemans calls it 'a lovely cottage-like building'[037])--is almost hidden in a rich profusion of roses and ivy and jessamine and virginia-creeper. Wordsworth, though he passionately admired the shapes and hues of flowers, knew nothing of their fragrance. In this respect knowledge at one entrance was quite shut out. He had possessed at no time of his life the sense of smell. To make up for this deficiency, he is said (by De Quincey) to have had "a peculiar depth of organic sensibility of form and color."

Mr. Justice Coleridge tells us that Wordsworth dealt with shrubs, flower-beds and lawns with the readiness of a practised landscape-gardener, and that it was curious to observe how he had imparted a portion of his taste to his servant, James Dixon. In fact, honest James regarded himself as a sort of Arbiter Elegantiarum. The master and his servant often discussed together a question of taste. Wordsworth communicated to Mr. Justice Coleridge how "he and James" were once "in a puzzle" about certain discolored spots upon the lawn. "Cover them with soap-lees," said the master. "That will make the green there darker than the rest," said the gardener. "Then we must cover the whole." "That will not do," objects the gardener, "with reference to the little lawn to which you pass from this." "Cover that," said the poet. "You will then," replied the gardener, "have an unpleasant contrast with the foliage surrounding it."

Pope too had communicated to his gardener at Twickenham something of his own taste. The man, long after his master's death, in reference to the training of the branches of plants, used to talk of their being made to hang *"something poetical"*.

It would have grieved Shakespeare and Pope and Shenstone had they anticipated the neglect or destruction of their beloved retreats. Wordsworth said, "I often ask myself what will become of Rydal Mount after our day. Will the old walls and steps remain in front of the house and about the grounds, or will they be swept away with all the beautiful mosses and ferns and wild geraniums and other flowers which their rude construction suffered and encouraged to grow among them. This little wild flower, *Poor Robin*, is here constantly courting my attention and exciting what may be called a domestic interest in the varying aspect of its stalks and leaves and flowers." I hope no Englishman meditating to reside on the grounds now sacred to the memory of a national poet will ever forget these words of the poet or treat his cottage and garden at Rydal Mount as some of Pope's countrymen have treated the house and grounds at Twickenham.[038] It would be sad indeed to hear, after this, that any one had refused to spare the *Poor Robins* and *wild geraniums* of Rydal Mount. Miss Jewsbury has a poem descriptive of "the Poet's Home." I must give the first stanza:--

WORDSWORTH'S COTTAGE.

> Low and white, yet scarcely seen Are its walls of mantling green; Not a window lets in light But through flowers clustering bright, Not a glance may wander there But it falls on something fair; Garden choice and fairy mound Only that no

elves are found; Winding walk and sheltered nook For student grave and graver book, Or a bird-like bower perchance Fit for maiden and romance.

Another lady-poet has poured forth in verse her admiration of THE RESIDENCE OF WORDSWORTH.

Not for the glory on their heads Those stately hill-tops wear, Although the summer sunset sheds Its constant crimson there: Not for the gleaming lights that break The purple of the twilight lake, Half dusky and half fair, Does that sweet valley seem to be A sacred place on earth to me. The influence of a moral spell Is found around the scene, Giving new shadows to the dell, New verdure to the green. With every mountain-top is wrought The presence of associate thought, A music that has been; Calling that loveliness to life, With which the inward world is rife. His home--our English poet's home-- Amid these hills is made; Here, with the morning, hath he come, There, with the night delayed. On all things is his memory cast, For every place wherein he past, Is with his mind arrayed, That, wandering in a summer hour, Asked wisdom of the leaf and flower.

L.E.L.

The cottage and garden of the poet are not only picturesque and delightful in themselves, but from their position in the midst of some of the finest scenery of England. One of the writers in the book entitled '*The Land we Live in*' observes that the bard of the mountains and the lakes could not have found a more fitting habitation had the whole land been before him, where to choose his place of rest. "Snugly sheltered by the mountains, embowered among trees, and having in itself prospects of surpassing beauty, it also lies in the midst of the very noblest objects in the district, and in one of the happiest social positions. The grounds are delightful in every respect; but one view-- that from the terrace of moss-like grass--is, to our thinking, the most exquisitely graceful in all this land of beauty. It embraces the whole valley of Windermere, with hills on either side softened into perfect loveliness."

Eustace, the Italian tourist, seems inclined to deprive the English of the honor of being the first cultivators of the natural style in gardening, and thinks that it was borrowed not from Milton but from Tasso. I suppose that most genuine poets, in all ages and in all countries, when they give full play to the imagination, have glimpses of the truly natural in the arts. The reader will probably be glad to renew his acquaintance with Tasso's description of the garden of Armida. I shall give the good old version of Edward Fairfax from the edition of 1687. Fairfax was a true poet and wrote musically at a time when sweetness of versification was not so much aimed at as in a later day. Waller confessed that he owed the smoothness of his verse to the example of Fairfax, who, as Warton observes, "well vowelled his lines."

THE GARDEN OF ARMIDA.

> When they had passed all those troubled ways, The Garden sweet spread forth her green to shew; The moving crystal from the fountains plays; Fair trees, high plants, strange herbs and flowerets new, Sunshiny hills, vales hid from Phoebus' rays, Groves, arbours, mossie caves at once they view, And that which beauty most, most wonder brought, No where appear'd the Art which all this wrought. So with the rude the polished mingled was, That natural seem'd all and every part, Nature would craft in counterfeiting pass, And imitate her imitator Art: Mild was the air, the skies were clear as glass, The trees no whirlwind felt, nor tempest's smart, But ere the fruit drop off, the blossom comes, This springs, that falls, that ripeneth and this blooms. The leaves upon the self-same bough did hide, Beside the young, the old and ripened fig, Here fruit was green, there ripe with vermeil side; The apples new and old grew on one twig, The fruitful vine her arms spread high and wide, That bended underneath their clusters big; The grapes were tender here, hard, young and sour, There purple ripe, and nectar sweet forth pour. The joyous birds, hid under green-wood shade, Sung merry notes on every branch and bow, The wind that in the leaves and waters plaid With murmur sweet, now sung and whistled now; Ceaséd the birds, the wind loud answer

made: And while they sung, it rumbled soft and low; Thus were it hap or cunning, chance or art, The wind in this strange musick bore his part. With party-coloured plumes and purple bill, A wondrous bird among the rest there flew, That in plain speech sung love-lays loud and shrill, Her leden was like humane language true; So much she talkt, and with such wit and skill, That strange it seeméd how much good she knew; Her feathered fellows all stood hush to hear, Dumb was the wind, the waters silent were. The gently budding rose (quoth she) behold, That first scant peeping forth with virgin beams, Half ope, half shut, her beauties doth upfold In their dear leaves, and less seen, fairer seems, And after spreads them forth more broad and bold, Then languisheth and dies in last extreams, Nor seems the same, that deckéd bed and bower Of many a lady late, and paramour. So, in the passing of a day, doth pass The bud and blossom of the life of man, Nor ere doth flourish more, but like the grass Cut down, becometh wither'd, pale and wan: O gather then the rose while time thou hast, Short is the day, done when it scant began; Gather the rose of love, while yet thou may'st Loving be lov'd; embracing, be embrac'd. He ceas'd, and as approving all he spoke, The quire of birds their heav'nly tunes renew, The turtles sigh'd, and sighs with kisses broke, The fowls to shades unseen, by pairs withdrew; It seem'd the laurel chaste, and stubborn oak, And all the gentle trees on earth that grew, It seem'd the land, the sea, and heav'n above, All breath'd out fancy sweet, and sigh'd out love.

Godfrey of Bulloigne

I must place near the garden of Armida, Ariosto's garden of Alcina. "Ariosto," says Leigh Hunt, "cared for none of the pleasures of the great, except building, and was content in Cowley's fashion, with "a small house in a large garden." He loved gardening better than he understood it, was always shifting his plants, and destroying the seeds, out of impatience to see them germinate. He was rejoicing once on the coming up of some "capers" which he had been visiting every day, to see how they got on, when it turned out that his capers were elder trees!"

THE GARDEN OF ALCINA.

> 'A more delightful place, wherever hurled, Through the whole air, Rogero had not found; And had he ranged the universal world, Would not have seen a lovelier in his round, Than that, where, wheeling wide, the courser furled His spreading wings, and lighted on the ground Mid cultivated plain, delicious hill, Moist meadow, shady bank, and crystal rill; 'Small thickets, with the scented laurel gay, Cedar, and orange, full of fruit and flower, Myrtle and palm, with interwoven spray, Pleached in mixed modes, all lovely, form a bower; And, breaking with their shade the scorching ray, Make a cool shelter from the noon-tide hour. And nightingales among those branches wing Their flight, and safely amorous descants sing. 'Amid red roses and white lilies *there*, Which the soft breezes freshen as they fly, Secure the cony haunts, and timid hare, And stag, with branching forehead broad and high. These, fearless of the hunter's dart or snare, Feed at their ease, or ruminating lie; While, swarming in those wilds, from tuft or steep, Dun deer or nimble goat disporting leap.'

Rose's Orlando Furioso.

Spenser's description of the garden of Adonis is too long to give entire, but I shall quote a few stanzas. The old story on which Spenser founds his description is told with many variations of circumstance and meaning; but we need not quit the pages of the Faerie Queene to lose ourselves amidst obscure mythologies. We have too much of these indeed even in Spenser's own version of the fable.

THE GARDEN OF ADONIS.

> Great enimy to it, and all the rest That in the Gardin of Adonis springs, Is wicked Time; who with his scythe addrest Does mow the flowring herbes and goodly things, And all their glory to the ground downe flings, Where they do wither and are fowly mard He flyes about, and with his flaggy wings Beates downe both leaves and buds without regard, Ne ever pitty may relent his malice hard.

But were it not that Time their troubler is, All that in this delightful gardin growes Should happy bee, and have immortall blis: For here all plenty and all pleasure flowes; And sweete Love gentle fitts emongst them throwes, Without fell rancor or fond gealosy. Franckly each paramour his leman knowes, Each bird his mate; ne any does envy Their goodly meriment and gay felicity. There is continual spring, and harvest there Continuall, both meeting at one tyme: For both the boughes doe laughing blossoms beare. And with fresh colours decke the wanton pryme, And eke attonce the heavy trees they clyme, Which seeme to labour under their fruites lode: The whiles the ioyous birdes make their pastyme Emongst the shady leaves, their sweet abode, And their trew loves without suspition tell abrode. Right in the middest of that Paradise There stood a stately mount, on whose round top A gloomy grove of mirtle trees did rise, Whose shady boughes sharp steele did never lop, Nor wicked beastes their tender buds did crop, But like a girlond compasséd the hight, And from their fruitfull sydes sweet gum did drop, That all the ground, with pretious deaw bedight, Threw forth most dainty odours and most sweet delight. And in the thickest covert of that shade There was a pleasaunt arber, not by art But of the trees owne inclination made, Which knitting their rancke braunches part to part, With wanton yvie-twine entrayld athwart, And eglantine and caprifole emong, Fashioned above within their inmost part, That neither Phoebus beams could through them throng, Nor Aeolus sharp blast could worke them any wrong. And all about grew every sort of flowre, To which sad lovers were transformde of yore, Fresh Hyacinthus, Phoebus paramoure And dearest love; Foolish Narcisse, that likes the watry shore; Sad Amaranthus, made a flowre but late, Sad Amaranthus, in whose purple gore Me seemes I see Amintas wretched fate, To whom sweet poet's verse hath given endlesse date.

Fairie Queene, Book III. Canto VI.

I must here give a few stanzas from Spenser's description of the *Bower of Bliss*

> In which whatever in this worldly state Is sweet and pleasing unto living sense, Or that may dayntiest fantasy aggrate Was pouréd forth with pleantiful dispence.

The English poet in his Fairie Queene has borrowed a great deal from Tasso and Ariosto, but generally speaking, his borrowings, like those of most true poets, are improvements upon the original.

THE BOWER OF BLISS.

> There the most daintie paradise on ground Itself doth offer to his sober eye, In which all pleasures plenteously abownd, And none does others happinesse envye; The painted flowres; the trees upshooting hye; The dales for shade; the hilles for breathing-space; The trembling groves; the christall running by; And that which all faire workes doth most aggrace, The art, which all that wrought, appearéd in no place. One would have thought, (so cunningly the rude[039] And scornéd partes were mingled with the fine,) That Nature had for wantonesse ensude Art, and that Art at Nature did repine; So striving each th' other to undermine, Each did the others worke more beautify; So diff'ring both in willes agreed in fine; So all agreed, through sweete diversity, This Gardin to adorn with all variety. And in the midst of all a fountaine stood, Of richest substance that on earth might bee, So pure and shiny that the silver flood Through every channel running one might see; Most goodly it with curious ymageree Was over-wrought, and shapes of naked boyes, Of which some seemed with lively iollitee To fly about, playing their wanton toyes, Whylest others did themselves embay in liquid ioyes.
>
> Eftsoones they heard a most melodious sound, Of all that mote delight a daintie eare, Such as attonce might not on living ground, Save in this paradise, be heard elsewhere: Right hard it was for wight which did it heare, To read what manner musicke that mote bee; For all that pleasing is to living eare Was there consorted in one harmonee; Birdes, voices, instruments, windes, waters all agree: The ioyous bir-

des, shrouded in chearefull shade, Their notes unto the voice attempred sweet; Th' angelicall soft trembling voyces made To th' instruments divine respondence meet; The silver-sounding instruments did meet With the base murmure of the waters fall; The waters fall with difference discreet, Now soft, now loud, unto the wind did call; The gentle warbling wind low answeréd to all.

The Faerie Queene, Book II. Canto XII.

Every school-boy has heard of the gardens of the Hesperides. The story is told in many different ways. According to some accounts, the Hesperides, the daughters of Hesperus, were appointed to keep charge of the tree of golden apples which Jupiter presented to Juno on their wedding day. A hundred-headed dragon that never slept, (the offspring of Typhon,) couched at the foot of the tree. It was one of the twelve labors of Hercules to obtain possession of some of these apples. He slew the dragon and gathered three golden apples. The gardens, according to some authorities, were situated near Mount Atlas.

Shakespeare seems to have taken *Hesperides* to be the name of the garden instead of that of its fair keepers. Even the learned Milton in his *Paradise Regained*, (Book II) talks of *the ladies of the Hesperides*, and appears to make the word Hesperides synonymous with "Hesperian gardens." Bishop Newton, in a foot-note to the passage in "Paradise Regained," asks, "What are the Hesperides famous for, but the gardens and orchards which *they had* bearing golden fruit in the western Isles of Africa." Perhaps after all there may be some good authority in favor of extending the names of the nymphs to the garden itself. Malone, while condemning Shakespeare's use of the words as inaccurate, acknowledges that other poets have used it in the same way, and quotes as an instance, the following lines from Robert Greene:--

Shew thee the tree, leaved with refined gold, Whereon the fearful dragon held his seat, That watched *the garden* called the *Hesperides*.

Robert Greene.

> For valour is not love a Hercules, Still climbing trees in the Hesperides?

Love's Labour Lost.

> Before thee stands this fair Hesperides, With golden fruit, but dangerous to be touched For death-like dragons here affright thee hard.

Pericles, Prince of Tyre.

Milton, after the fourth line of his Comus, had originally inserted, in his manuscript draft of the poem, the following description of the garden of the Hesperides.

THE GARDEN OF THE HESPERIDES

> Amid the Hesperian gardens, on whose banks Bedewed with nectar and celestial songs Eternal roses grow, and hyacinth, And fruits of golden rind, on whose fair tree The scaly harnessed dragon ever keeps His uninchanted eye, around the verge And sacred limits of this blissful Isle The jealous ocean that old river winds His far extended aims, till with steep fall Half his waste flood the wide Atlantic fills; And half the slow unfathomed Stygian pool But soft, I was not sent to court your wonder With distant worlds and strange removéd climes Yet thence I come and oft from thence behold The smoke and stir of this dim narrow spot

Milton subsequently drew his pen through these lines, for what reason is not known. Bishop Newton observes, that this passage, saved from intended destruction, may serve as a specimen of the truth of the observation that

> Poets lose half the praise they should have got Could it be known what they discreetly blot.

Waller.

As I have quoted in an earlier page some unfavorable allusions to Homer's description of a Grecian garden, it will be but fair to follow up Milton's picture of Paradise, and Tasso's garden of Armida, and Ariosto's Garden of Alcina, and Spenser's Garden of Adonis and his Bower of Bliss, with Homer's description of the Garden of Alcinous. Minerva tells Ulysses that the Royal mansion to which the garden of Alcinous is attached is of such conspicuous grandeur and so generally known, that any child might lead him to it;

> For Phoeacia's sons Possess not houses equalling in aught
> The mansion of Alcinous, the king.

I shall give Cowper's version, because it may be less familiar to the reader than Pope's, which is in every one's hand.

THE GARDEN OF ALCINOUS

> Without the court, and to the gates adjoined A spacious garden lay, fenced all around, Secure, four acres measuring complete, There grew luxuriant many a lofty tree, Pomgranate, pear, the apple blushing bright, The honeyed fig, and unctuous olive smooth. Those fruits, nor winter's cold nor summer's heat Fear ever, fail not, wither not, but hang Perennial, while unceasing zephyr breathes Gently on all, enlarging these, and those Maturing genial; in an endless course. Pears after pears to full dimensions swell, Figs follow figs, grapes clustering grow again Where clusters grew, and (every apple stripped) The boughs soon tempt the gatherer as before. There too, well rooted, and of fruit profuse, His vineyard grows; part, wide extended, basks In the sun's beams; the arid level glows; In part they gather, and in part they tread The wine-press, while, before the eye, the grapes Here put their blossoms forth, there gather fast Their blackness. On the garden's verge extreme Flowers of all hues[040] smile all the year, arranged With neatest art judicious, and amid The lovely scene two fountains welling forth, One visits, into every part diffused, The garden-ground, the other soft beneath The threshold steals into the palace court Whence every citizen his vase supplies.

Homer's Odyssey, Book VII.

The mode of watering the garden-ground, and the use made of the water by the public--

Whence every citizen his vase supplies--

can hardly fail to remind Indian and Anglo-Indian readers of a Hindu gentleman's garden in Bengal.

Pope first published in the *Guardian* his own version of the account of the garden of Alcinous and subsequently gave it a place in his entire translation of Homer. In introducing the readers of the *Guardian* to the garden of Alcinous he observes that "the two most celebrated wits of the world have each left us a particular picture of a garden; wherein those great masters, being wholly unconfined and pointing at pleasure, may be thought to have given a full idea of what seemed most excellent in that way. These (one may observe) consist entirely of the useful part of horticulture, fruit trees, herbs, waters, &c. The pieces I am speaking of are Virgil's account of the garden of the old Corycian, and Homer's of that of Alcinous. The first of these is already known to the English reader, by the excellent versions of Mr. Dryden and Mr. Addison."

I do not think our present landscape-gardeners, or parterre-gardeners or even our fruit or kitchen-gardeners can be much enchanted with Virgil's ideal of a garden, but here it is, as "done into English," by John Dryden, who describes the Roman Poet as "a profound naturalist," and "*a curious Florist.*"

THE GARDEN OF THE OLD CORYCIAN.

> I chanc'd an old Corycian swain to know, Lord of few acres, and those barren too, Unfit for sheep or vines, and more unfit to sow: Yet, lab'ring well his little spot of ground, Some scatt'ring pot-herbs here and there he found, Which, cultivated with his daily care And bruis'd with vervain, were his frugal fare. With wholesome poppy-flow'rs, to mend his homely board: For, late returning home, he supp'd at ease, And wisely deem'd the wealth of monarchs less: The little of his own, because his own, did please. To quit his care, he ga-

ther'd, first of all, In spring the roses, apples in the fall: And, when cold winter split the rocks in twain, And ice the running rivers did restrain, He stripp'd the bear's foot of its leafy growth, And, calling western winds, accus'd the spring of sloth He therefore first among the swains was found To reap the product of his labour'd ground, And squeeze the combs with golden liquor crown'd His limes were first in flow'rs, his lofty pines, With friendly shade, secur'd his tender vines. For ev'ry bloom his trees in spring afford, An autumn apple was by tale restor'd He knew to rank his elms in even rows, For fruit the grafted pear tree to dispose, And tame to plums the sourness of the sloes With spreading planes he made a cool retreat, To shade good fellows from the summer's heat

Virgil's *Georgics, Book IV*.

An excellent Scottish poet--Allan Ramsay--a true and unaffected describer of rural life and scenery--seems to have had as great a dislike to topiary gardens, and quite as earnest a love of nature, as any of the best Italian poets. The author of the "Gentle Shepherd" tells us in the following lines what sort of garden most pleased his fancy.

ALLAN RAMSAY'S GARDEN.

I love the garden wild and wide, Where oaks have plum-trees by their side, Where woodbines and the twisting vine Clip round the pear tree and the pine Where mixed jonquils and gowans grow And roses midst rank clover grow Upon a bank of a clear strand, In wrimplings made by Nature's hand Though docks and brambles here and there May sometimes cheat the gardener's care, *Yet this to me is Paradise, Compared with prim cut plots and nice, Where Nature has to Act resigned, Till all looks mean, stiff and confined.*

I cannot say that I should wish to see forest trees and docks and brambles in garden borders. Honest Allan here runs a little into the extreme, as men are apt enough to do, when they try to get as far as possible from the side advocated by an opposite party.

I shall now exhibit two paintings of bowers. I begin with one from Spenser.

A BOWER

> And over him Art stryving to compayre With Nature did an arber greene dispied[041] Framéd of wanton yvie, flouring, fayre, Through which the fragrant eglantine did spred His prickling armes, entrayld with roses red, Which daintie odours round about them threw And all within with flowers was garnishéd That, when myld Zephyrus emongst them blew, Did breathe out bounteous smels, and painted colors shew And fast beside these trickled softly downe A gentle streame, whose murmuring wave did play Emongst the pumy stones, and made a sowne, To lull him soft asleepe that by it lay The wearie traveiler wandring that way, Therein did often quench his thirsty head And then by it his wearie limbes display, (Whiles creeping slomber made him to forget His former payne,) and wypt away his toilsom sweat. And on the other syde a pleasaunt grove Was shott up high, full of the stately tree That dedicated is t'Olympick Iove, And to his son Alcides,[042] whenas hee In Nemus gaynéd goodly victoree Theirin the merry birds of every sorte Chaunted alowd their cheerful harmonee, And made emongst themselves a sweete consórt That quickned the dull spright with musicall comfórt.

Fairie Queene, Book 2 Cant. 5 Stanzas 29, 30 and 31.

Here is a sweet picture of a "shady lodge" from the hand of Milton.

EVE'S NUPTIAL BOWER.

> Thus talking, hand in hand alone they pass'd On to their blissful bower. It was a place Chosen by the sov'reign Planter, when he framed All things to man's delightful use, the roof Of thickest covert was inwoven shade, Laurel and myrtle, and what higher grew Of firm and fragrant leaf, on either side Acanthus, and each odorous bushy shrub, Fenced up the

> verdant wall, each beauteous flower Iris all hues, roses, and jessamine, Rear'd high their flourish'd heads between, and wrought Mosaic, under foot the violet, Crocus, and hyacinth, with rich inlay Broider'd the ground, more colour'd than with stone Of costliest emblem other creature here, Beast, bird, insect, or worm, durst enter none, Such was their awe of man. In shadier bower More sacred and sequester'd, though but feign'd, Pan or Sylvanus never slept, nor nymph Nor Faunus haunted. Here, in close recess, With flowers, garlands, and sweet smelling herbs, Espoused Eve deck'd first her nuptial bed, And heavenly quires the hymenean sung

I have already quoted from Leigh Hunt's "Stories from the Italian poets" an amusing anecdote illustrative of Ariosto's ignorance of botany. But even in these days when all sorts of sciences are forced upon all sorts of students, we often meet with persons of considerable sagacity and much information of a different kind who are marvellously ignorant of the vegetable world.

In the just published Memoirs of the late James Montgomery, of Sheffield, it is recorded that the poet and his brother Robert, a tradesman at Woolwich, (not Robert Montgomery, the author of 'Satan,' &c.) were one day walking together, when the trader seeing a field of flax in full flower, asked the poet what sort of corn it was. "Such corn as your shirt is made of," was the reply. "But Robert," observes a writer in the *Athenaeum*, "need not be ashamed of his simplicity. Rousseau, naturalist as he was, could hardly tell one berry from another, and three of our greatest wits disputing in the field whether the crop growing there was rye, barley, or oats, were set right by a clown, who truly pronounced it wheat."

Men of genius who have concentrated all their powers on some one favorite profession or pursuit are often thus triumphed over by the vulgar, whose eyes are more observant of the familiar objects and details of daily life and of the scenes around them. Wordsworth and Coleridge, on one occasion, after a long drive, and in the absence of a groom, endeavored to relieve the tired horse of its harness. After torturing the poor animal's neck and endangering its eyes by their clumsy and vain attempts to slip off the collar, they at last gave up the matter in despair. They felt convinced that the horse's head must have swollen since the collar was put on. At last a servant-girl beheld their perplexity. "La, masters," she exclaimed, "you dont set about it the right way." She then seized hold of the collar, turned it broad end up, and slipped it off in a second. The mystery that had puzzled two of the finest intellects of their time was a very simple matter indeed to a country wench who had perhaps never heard that England possessed a Shakespeare.

James Montgomery was a great lover of flowers, and few of our English poets have written about the family of Flora, the sweet wife of Zephyr, in a more genial spirit. He used to regret that the old Floral games and processions on May-day and other holidays had gone out of fashion. Southey tells us that in George the First's reign a grand Florist's Feast was held at Bethnall Green, and that a carnation named after his Majesty was *King of the Year*. The Stewards were dressed with laurel leaves and flowers. They carried gilded staves. Ninety cultivators followed in procession to the sound of music, each bearing his own flowers before him. All elegant customs of this nature have fallen into desuetude in England, though many of them are still kept up in other parts of Europe.

Chaucer who dearly loved all images associated with the open air and the dewy fields and bright mornings and radiant flowers makes the gentle Emily,

> That fairer was to seene Than is the lily upon his stalkie greene,

rise early and do honor to the birth of May-day. All things now seem to breathe of hope and joy.

> Though long hath been The trance of Nature on the naked bier Where ruthless Winter mocked her slumbers drear And rent with icy hand her robes of green, That trance is brightly broken! Glossy trees, Resplendent meads and variegated flowers Flash in the sun and flutter in the breeze And now with dreaming eye the poet sees Fair shapes of pleasure haunt romantic bowers, And laughing streamlets chase the flying hours.

D.L.R.

The great describer of our Lost Paradise did not disdain to sing a SONG ON MAY-MORNING.

> Now the bright Morning star, Day's harbinger, Comes dancing from the east, and leads with her The flowery May, who from her green lap throws The yellow cowslip and the pale primrose Hail bounteous-May, that dost inspire Mirth and youth and warm desire; Woods and groves are of thy dressing, Hill and dale do boast thy blessing. Thus we salute thee with our early song, And welcome thee and wish thee long.

Nor did the Poet of the World, William Shakespeare, hesitate to

> Do observance to a morn of May.

He makes one of his characters (in *King Henry VIII.*) complain that it is as impossible to keep certain persons quiet on an ordinary day, as it is to make them sleep on May-day--once the time of universal merriment-- when every one was wont "*to put himself into triumph.*"

> 'Tis as much impossible, Unless we sweep 'em from the doors with cannons To scatter 'em, *as 'tis to make 'em sleep On May-day Morning.*

Spenser duly celebrates, in his "Shepheard's Calender,"

> Thilke mery moneth of May When love-lads masken in fresh aray,

when "all is yclad with pleasaunce, the ground with grasse, the woods with greene leaves, and the bushes with bloosming buds."

> Sicker[043] this morowe, no longer agoe, I saw a shole of shepeardes outgoe With singing and shouting and iolly chere: Before them yode[044] a lustre tabrere,[045] That to the many a hornepype playd Whereto they dauncen eche one with his mayd. To see those folks make such iovysaunce,

Made my heart after the pype to daunce. Tho[046] to the greene wood they speeden hem all To fetchen home May with their musicall; And home they bringen in a royall throne Crowned as king; and his queene attone[047] Was LADY FLORA.

Spenser.

This is the season when the birds seem almost intoxicated with delight at the departure of the dismal and cold and cloudy days of winter and the return of the warm sun. The music of these little May musicians seems as fresh as the fragrance of the flowers. The Skylark is the prince of British Singing-birds--the leader of their cheerful band.

LINES TO A SKYLARK.

Wanderer through the wilds of air! Freely as an angel fair Thou dost leave the solid earth, Man is bound to from his birth Scarce a cubit from the grass Springs the foot of lightest lass-- *Thou* upon a cloud can'st leap, And o'er broadest rivers sweep, Climb up heaven's steepest height, Fluttering, twinkling, in the light, Soaring, singing, till, sweet bird, Thou art neither seen nor heard, Lost in azure fields afar Like a distance hidden star, That alone for angels bright Breathes its music, sheds its light Warbler of the morning's mirth! When the gray mists rise from earth, And the round dews on each spray Glitter in the golden ray, And thy wild notes, sweet though high, Fill the wide cerulean, sky, Is there human heart or brain Can resist thy merry strain? But not always soaring high, Making man up turn his eye Just to learn what shape of love, Raineth music from above,-- All the sunny cloudlets fair Floating on the azure air, All the glories of the sky Thou leavest unreluctantly, Silently with happy breast To drop into thy lowly nest. Though the frame of man must be Bound to earth, the soul is free, But that freedom oft doth bring Discontent and sorrowing. Oh! that from each waking vision, Gorgeous vista, gleam Elysian, From ambition's dizzy height, And from hope's illusive light, Man, like thee, glad lark, could brook Upon a low green spot to look, And with home affections blest Sink into as calm a nest! D.L.R.

I brought from England to India two English skylarks. I thought they would help to remind me of English meadows and keep alive many agreeable home-associations. In crossing the desert they were carefully lashed on the top of one of the vans, and in spite of the dreadful jolting and the heat of the sun they sang the whole way until night-fall. It was pleasant to hear English larks from rich clover fields singing so joyously in the sandy waste. In crossing some fields between Cairo and the Pyramids I was surprized and delighted with the songs of Egyptian skylarks. Their notes were much the same as those of the English lark. The lark of Bengal is about the size of a sparrow and has a poor weak note. At this moment a lark from Caubul (larger than an English lark) is doing his best to cheer me with his music. This noble bird, though so far from his native fields, and shut up in his narrow prison, pours forth his rapturous melody in an almost unbroken stream from dawn to sunset. He allows no change of season to abate his minstrelsy, to any observable degree, and seems equally happy and musical all the year round. I have had him nearly two years, and though of course he must moult his feathers yearly, I have not observed the change of plumage, nor have I noticed that he has sung less at one period of the year than another. One of my two English larks was stolen the very day I landed in India, and the other soon died. The loss of an English lark is not to be replaced in Calcutta, though almost every week, canaries, linnets, gold- finches and bull-finches are sold at public auctions here.

But I must return to my main subject.--The ancients used to keep the great Feast of the goddess Flora on the 28th of April. It lasted till the 3rd of May. The Floral Games of antiquity were unhappily debased by indecent exhibitions; but they were not entirely devoid of better characteristics.[048] Ovid describing the goddess Flora says that "while she was speaking she breathed forth vernal roses from her mouth." The same poet has represented her in her garden with the Florae gathering flowers and the Graces making garlands of them. The British borrowed the idea of this festival from the Romans. Some of our Kings and Queens used '*to go a Maying*,' and to have feasts of wine and venison in the open meadows or under the good green-wood. Prior says:

> Let one great day To celebrate sports and floral play Be set aside.

But few people, in England, in these times, distinguish May-day from the initial day of any other month of the twelve. I am old enough to remember *Jack-in-the-Green*. Nor have I forgotten the cheerful clatter--the brush-and-shovel music--of our little British negroes--"innocent blacknesses," as Lamb calls them--the chimney-sweepers,--a class now almost *swept away* themselves by *machinery*. One May-morning in the streets of London these tinsel-decorated merry- makers with their sooty cheeks and black lips lined with red, and staring eyes whose white seemed whiter still by contrast with the darkness of their cases, and their ivory teeth kept sound and brilliant with the professional powder, besieged George Selwyn and his arm-in-arm companion, Lord Pembroke, for May-day boxes. Selwyn making them a low bow, said, very solemnly "I have often heard of *the sovereignty of the people*, and I suppose you are some of the young princes in court mourning."

My Native readers in Bengal can form no conception of the delight with which the British people at home still hail the spring of the year, or the deep interest which they take in all "the Seasons and their change"; though they have dropped some of the oldest and most romantic of the ceremonies once connected with them. If there were an annual fall of the leaf in the groves of India, instead of an eternal summer, the natives would discover how much the charms of the vegetable world are enhanced by these vicissitudes, and how even winter itself can be made delightful. My brother exiles will remember as long as life is in them, how exquisite, in dear old England, is the enjoyment of a brisk morning walk in the clear frosty air, and how cheering and cosy is the social evening fire! Though a cold day in Calcutta is not exactly like a cold day in London, it sometimes revives the remembrance of it. An Indian winter, if winter it may be called, is indeed far less agreeable than a winter in England, but it is not wholly without its pleasures. It is, at all events, a grateful change--a welcome relief and refreshment after a sultry summer or a *muggy* rainy season.

An Englishman, however, must always prefer the keener but more wholesome frigidity of his own clime. There, the external gloom and bleakness of a severe winter day enhance our in-door comforts, and we do not miss sunny skies when greeted with sunny looks. If we then see no blooming flowers, we see blooming faces. But as we have few domestic enjoyments in this country--no social snugness,--no sweet seclusion--and as our houses are as open as bird-cages,--and as we almost live in public and in the open air--we have little comfort when compelled, with an enfeebled frame and a morbidly sensitive cuticle, to remain at home on what an Anglo-Indian Invalid calls a cold day, with an easterly wind whistling through every room.[049] In our dear native country each season has its peculiar moral or physical attractions. It is not easy to say which is the most agreeable--its summer or its winter. Perhaps I must decide in favor of the first. The memory of many a smiling summer day still flashes upon my soul. If the whole of human life were like a fine English day in June, we should cease to wish for "another and a better world." It is often from dawn to sunset one revel of delight. How pleasantly, from the first break of day, have I lain wide awake and traced the approach of the breakfast hour by the increasing notes of birds and the advancing sun- light on my curtains! A summer feeling, at such a time, would make my heart dance within me, as I thought of the long, cheerful day to be enjoyed, and planned some rural walk, or rustic entertainment. The ills that flesh is heir to, if they occurred for a moment, appeared like idle visions. They were inconceivable as real things. As I heard the lark singing in "a glorious privacy of light," and saw the boughs of the green and gold laburnum waving at my window, and had my fancy filled with images of natural beauty, I felt a glow of fresh life in my veins, and my soul was inebriated with joy. It is difficult, amidst such exhilarating influences, to entertain those melancholy ideas which sometimes crowd upon, us, and appear so natural, at a less happy hour. Even actual misfortune comes in a questionable shape, when our physical constitution is in perfect health, and the flowers are in full bloom, and the skies are blue, and the streams are glittering in the sun. So powerfully does the light of external nature sometimes act upon the moral system, that a sweet sensation steals gradually over the heart, even when we think we have reason to be sorrowful, and while we almost accuse ourselves of a want of feeling. The fretful hypochondriac would do well to bear this fact in mind, and not take it for granted that all are cold and selfish who fail to sympathize with his fantastic cares. He should remember that men are sometimes so buoyed up by the sense of corporeal power, and a communion with nature in her cheerful moods, that things connected with their own personal interests, and which at other times might irritate and wound their feelings, pass by them like the idle wind which they regard not. He himself must have had his intervals of comparative happiness, in which the causes of his present grief would have appeared trivial and absurd. He should not, then, expect persons whose blood is warm in their veins, and whose eyes are open to the blessed sun in heaven, to think more of the apparent causes of his sorrow than he would himself, were his mind and body in a healthful state.

With what a light heart and eager appetite did I enter the little breakfast parlour of which the glass-doors opened upon a bright green lawn, variegated with small beds of flowers! The table was spread with dewy and delicious fruits from our own garden, and gathered by fair and friendly hands. Beautiful and luscious as were these garden dainties, they were of small account in comparison with the fresh cheeks and cherry lips that so frankly accepted the wonted early greeting. Alas! how that circle of early friends is now divided, and what a change has since come over the spirit of our dreams! Yet still I cherish boyish feelings, and the past is sometimes present. As I give an imaginary kiss to an "old familiar face," and catch myself almost unconsciously, yet literally, returning imaginary smiles, my heart is as fresh and fervid as of yore.

A lapse of fifteen years, and a distance of fifteen thousand miles, and the glare of a tropical sky and the presence of foreign faces, need not make an Indian Exile quite forgetful of home-delights. Parted friends may still share the light of love as severed clouds are equally kindled by the same sun. No number of miles or days can change or separate faithful spirits or annihilate early associations. That strange magician, Fancy, who supplies so many corporeal deficiencies and overcomes so many physical obstructions, and mocks at space and time, enables us to pass in the twinkling of an eye over the dreary waste of waters that separates the exile from the scenes and companions of his youth. He treads again his native shore. He sits by the hospitable hearth and listens to the ringing laugh of children. He exchanges cordial greetings with the "old familiar faces." There is a resurrection of the dead, and a return of vanished years. He abandons himself to the sweet illusion, and again

> Lives over each scene, and is what he beholds.

I must not be too egotistically garrulous in print, or I would now attempt to describe the various ways in which I have spent a summer's day in England. I would dilate upon my noon-day loiterings amidst wild ruins, and thick forests, and on the shaded banks of rivers--the pic-nic parties--the gipsy prophecies--the twilight homeward walk--the social tea-drinking, and, the last scene of all, the "rosy dreams and slumbers light," induced by wholesome exercise and placid thoughts.[050] But perhaps these few simple allusions are sufficient to awaken a train of kindred associations in the reader's mind, and he will thank me for those words and images that are like the keys of memory, and "open all her cells with easy force."

If a summer's day be thus rife with pleasure, scarcely less so is a day in winter, though with some little drawbacks, that give, by contrast, a zest to its enjoyments. It is difficult to leave the warm morning bed and brave the external air. The fireless grate and frosted windows may well make the stoutest shudder. But when we have once screwed our courage to the sticking place, and with a single jerk of the clothes, and a brisk jump from the bed, have commenced the operations of the toilet, the battle is nearly over. The teeth chatter for a while, and the limbs shiver, and we do not feel particularly comfortable while breaking the ice in our jugs, and performing our cold ablutions amidst the sharp, glass-like fragments, and wiping our faces with a frozen towel. But these petty evils are quickly vanquished, and as we rush out of the house, and tread briskly and firmly on the hard ringing earth, and breathe our visible breath in the clear air, our strength and self- importance miraculously increase, and the whole frame begins to glow. The warmth and vigour thus acquired are inexpressibly delightful. As we re-enter the house, we are proud of our intrepidity and vigour, and pity the effeminacy of our less enterprising friends, who, though huddled together round the fire, like flies upon a sunny wall, still complain of cold, and instead of the bloom of health and animation, exhibit pale and pinched and discolored features, and hands cold, rigid, and of a deadly hue. Those who rise with spirit on a winter morning, and stir and thrill themselves with early exercise, are indifferent to the cold for the rest of the day, and feel a confidence in their corporeal energies, and a lightness of heart that are experienced at no other season.

But even the timid and luxurious are not without their pleasures. As the shades of evening draw in, the parlour twilight--the closed curtains-- and the cheerful fire--make home a little paradise to all.

> Now stir the fire, and close the shutters fast, Let fall the curtains, wheel the sofa round, And while the bubbling and loud hissing urn Throws up a steamy column, and the cups That cheer but not inebriate wait on each, So let us welcome peaceful evening in

Cowper.

The warm and cold seasons of India have no charms like those of England, but yet people who are guiltless of what Milton so finely calls "a sullenness against nature," and who are willing, in a spirit of true philosophy and piety, to extract good from every thing, may save themselves from wretchedness even in this land of exile. While I am writing this paragraph, a bird in my room, (not the Caubul songster that I have already alluded to, but a fine little English linnet,) who is as much a foreigner here as I am, is pouring out his soul in a flood of song. His notes ring with joy. He pines not for his native meadows--he cares not for his wiry bars--he envies not the little denizens of air that sometimes flutter past my window, nor imagines, for a moment, that they come to mock him with their freedom. He is contented with his present enjoyments, because they are utterly undisturbed by idle comparisons with those experienced in the past or anticipated in the future. He has no thankless repinings and no vain desires. Is intellect or reason then so fatal, though sublime a gift that we cannot possess it without the poisonous alloy of care? Must grief and ingratitude inevitably find entrance into the heart, in proportion to the loftiness and number of our mental endowments? Are we to seek for happiness in ignorance? To these questions the reply is obvious. Every good quality may be abused, and the greatest, most; and he who perversely employs his powers of thought and imagination to a wrong purpose deserves the misery that he gains. Were we honestly to deduct from the ills of life all those of our own creation, how trifling, in the majority of cases, the amount that would remain! We seem to invite and encourage sorrow, while happiness is, as it were, forced upon us against our will. It is wonderful how some men pertinaciously cling to care, and argue themselves into a dissatisfaction with their lot. Thus it is really a matter of little moment whether fortune smile or frown, for it is in vain to look for superior felicity amongst those who have more "appliances and means to boot," than their fellow-men. Wealth, rank, and reputation, do not secure their possessors from the misery of discontent.

As happiness then depends upon the right direction and employment of our faculties, and not on worldly goods or mere localities, our countrymen might be cheerful enough, even in this foreign land, if they would only accustom themselves to a proper train of thinking, and be ready on every occasion to look on the brighter side of all things.[051] In reverting to home-scenes we should regard them for their intrinsic charms, and not turn them into a source of disquiet by mournfully comparing them with those around us. India, let Englishmen murmur as they will, has some attractions, enjoyments and advantages. No Englishman is here in danger of dying of starvation as some of our poets have done in the inhospitable streets of London. The comparatively princely and generous style in which we live in this country, the frank and familiar tone of our little society, and the general mildness of the climate, (excepting a few months of a too sultry summer) can hardly be denied by the most determined malcontent. The weather is indeed too often a great deal warmer than we like it; but if "the excessive heat" did not form a convenient subject for complaint and conversation, it is perhaps doubtful if it would so often be thought of or alluded to. But admit the objection. What climate is without its peculiar evils?

In the cold season a walk in India either in the morning or the evening is often extremely pleasant in pleasant company, and I am glad to see many sensible people paying the climate the compliment of treating it like that of England. It is now fashionable to use our limbs in the ordinary way, and the "Garden of Eden"[052] has become a favorite promenade, particularly on the evenings when a band from the Fort fills the air with a cheerful harmony and throws a fresher life upon the scene. It is not to be denied that besides the mere exercise, pedestrians at home have great advantages over those who are too indolent or aristocratic to leave their equipages, because they can cut across green and quiet fields, enter rural by-ways, and enjoy a thousand little patches of lovely scenery that are secrets to the high-road traveller. But still the Calcutta pedestrian has also his gratifications. He can enjoy no exclusive prospects, but he beholds upon an Indian river a forest of British masts--the noble shipping of the Queen of the Sea--and has a fine panoramic view of this City of Palaces erected by his countrymen on a foreign shore;-- and if he is fond of children, he must be delighted with the numberless pretty and happy little faces--the fair forms of Saxon men and women in miniature--that crowd about him on the green sward;--he must be charmed with their innocent prattle, their quick and graceful movements, and their winning ways, that awaken a tone of tender sentiment in his heart, and rekindle many sweet associations.

SONNETS,

WRITTEN IN EXILE.

> I. Man's heart may change, but Nature's glory never;-- And while the soul's internal cell is bright, The cloudless eye lets in the bloom and light Of earth and heaven to charm and cheer us ever. Though youth hath vanished, like a winding river Lost in the shadowy woods; and the dear sight Of native hill and nest-like cottage white, 'Mid breeze-stirred boughs whose crisp leaves gleam and quiver, And murmur sea-like sounds, perchance no more My homeward step shall hasten cheerily; Yet still I feel as I have felt of yore, And love this radiant world. Yon clear blue sky-- These gorgeous groves--this flower-enamelled floor-- Have deep enchantments for my heart and eye. II. Man's heart may change, but Na-

ture's glory never, Though to the sullen gaze of grief the sight Of sun illumined skies may *seem* less bright, Or gathering clouds less grand, yet she, as ever, Is lovely or majestic. Though fate sever The long linked bands of love, and all delight Be lost, as in a sudden starless night, The radiance may return, if He, the giver Of peace on earth, vouchsafe the storm to still This breast once shaken with the strife of care Is touched with silent joy. The cot--the hill, Beyond the broad blue wave--and faces fair, Are pictured in my dreams, yet scenes that fill My waking eye can save me from despair. III. Man's heart may change, but Nature's glory never,-- Strange features throng around me, and the shore Is not my own dear land. Yet why deplore This change of doom? All mortal ties must sever. The pang is past,--and now with blest endeavour I check the ready tear, the rising sigh The common earth is here--the common sky-- The common FATHER. And how high soever O'er other tribes proud England's hosts may seem, God's children, fair or sable, equal find A FATHER'S love. Then learn, O man, to deem All difference idle save of heart or mind Thy duty, love--each cause of strife, a dream-- Thy home, the world--thy family, mankind.

D.L.R.

For the sake of my home readers I must now say a word or two on the effect produced upon the mind of a stranger on his approach to Calcutta from the Sandheads.

As we run up the Bay of Bengal and approach the dangerous Sandheads, the beautiful deep blue of the ocean suddenly disappears. It turns into a pale green. The sea, even in calm weather, rolls over soundings in long swells. The hue of the water is varied by different depths, and in passing over the edge of soundings, it is curious to observe how distinctly the form of the sands may be traced by the different shades of green in the water above and beyond them. In the lower part of the bay, the crisp foam of the dark sea at night is instinct with phosphoric lustre. The ship seems to make her way through galaxies of little ocean stars. We lose sight of this poetical phenomenon as we approach the mouth of the Hooghly. But the passengers, towards the termination of their voyage, become less observant of the changeful aspect of the sea. Though amused occasionally by flights of sea-gulls, immense shoals of porpoises, apparently tumbling or rolling head over tail against the wind, and the small sprat-like fishes that sometimes play and glitter on the surface, the stranger grows impatient to catch a glimpse of an Indian jungle; and even the swampy tiger-haunted Saugor Island is greeted with that degree of interest which novelty usually inspires.

At first the land is but little above the level of the water. It rises gradually as we pass up further from the sea. As we come still nearer to Calcutta, the soil on shore seems to improve in richness and the trees to increase in size. The little clusters of nest-like villages snugly sheltered in foliage--the groups of dark figures in white garments--the cattle wandering over the open plain--the emerald-colored fields of rice--the rich groves of mangoe trees--the vast and magnificent banyans, with straight roots dropping from their highest branches, (hundreds of these branch-dropped roots being fixed into the earth and forming "a pillared shade"),--the tall, slim palms of different characters and with crowns of different forms, feathery or fan-like,--the many-stemmed and long, sharp-leaved bamboos, whose thin pliant branches swing gracefully under the weight of the lightest bird,--the beautifully rounded and bright green peepuls, with their burnished leaves glittering in the sunshine, and trembling at the zephyr's softest touch with a pleasant rustling sound, suggestive of images of coolness and repose,--form a striking and singularly interesting scene (or rather succession of scenes) after the monotony of a long voyage during which nothing has been visible but sea and sky.

But it is not until he arrives at a bend of the river called *Garden Reach*, where the City of Palaces first opens on the view, that the stranger has a full sense of the value of our possessions in the East. The princely mansions on our right;--(residences of English gentry), with their rich gardens and smooth slopes verdant to the water's edge,-- the large and rich Botanic Garden and the Gothic edifice of Bishop's College on our left--and in front, as we advance a little further, the countless masts of vessels of all sizes and characters, and from almost every clime,--Fort William, with its grassy ramparts and white barracks,--the Government House, a magnificent edifice in spite of many imperfections,--the substantial looking

Town Hall--the Supreme Court House--the broad and ever verdant plain (or *madaun*) in front--and the noble lines of buildings along the Esplanade and Chowringhee Road,--the new Cathedral almost at the extremity of the plain, and half-hidden amidst the trees,--the suburban groves and buildings of Kidderpore beyond, their outlines softened by the haze of distance, like scenes contemplated through colored glass--the high-sterned budgerows and small trim bauleahs along the edge of the river,--the neatly-painted palanquins and other vehicles of all sorts and sizes,--the variously- hued and variously-clad people of all conditions; the fair European, the black and nearly naked Cooly, the clean-robed and lighter-skinned native Baboo, the Oriental nobleman with his jewelled turban and kincob vest, and costly necklace and twisted cummerbund, on a horse fantastically caparisoned, and followed in barbaric state by a train of attendants with long, golden-handled punkahs, peacock feather chowries, and golden chattahs and silver sticks,--present altogether a scene that is calculated to at once delight and bewilder the traveller, to whom all the strange objects before him have something of the enchantment and confusion of an Arabian Night's dream. When he recovers from his surprise, the first emotion in the breast of an Englishman is a feeling of national pride. He exults in the recognition of so many glorious indications of the power of a small and remote nation that has founded a splendid empire in so strange and vast a land.

When the first impression begins to fade, and he takes a closer view of the great metropolis of India--and observes what miserable straw huts are intermingled with magnificent palaces--how much Oriental filth and squalor and idleness and superstition and poverty and ignorance are associated with savage splendour, and are brought into immediate and most incongruous contact with Saxon energy and enterprize and taste and skill and love of order, and the amazing intelligence of the West in this nineteenth century--and when familiarity breeds something like contempt for many things that originally excited a vague and pleasing wonder--the English traveller in the East is apt to dwell too exclusively on the worst side of the picture, and to become insensible to the real interest, and blind to the actual beauty of much of the scene around him. Extravagant astonishment and admiration, under the influence of novelty, a strong re-action, and a subsequent feeling of unreasonable disappointment, seem, in some degree, natural to all men; but in no other part of the world, and under no other circumstances, is this peculiarity of our condition more conspicuously displayed than in the case of Englishmen in India. John Bull, who is always a grumbler even on his own shores, is sure to become a still more inveterate grumbler in other countries, and perhaps the climate of Bengal, producing lassitude and low spirits, and a yearning for their native land, of which they are so justly proud, contribute to make our countrymen in the East even more than usually unsusceptible of pleasurable emotions until at last they turn away in positive disgust from the scenes and objects which remind them that they are in a state of exile.

"There is nothing," says Hamlet, "either good or bad, but thinking makes it so." At every change of the mind's colored optics the scene before it changes also. I have sometimes contemplated the vast metropolis of England--or rather *of the world*--multitudinous and mighty LONDON--with the pride and hope and exultation, not of a patriot only, but of a cosmopolite--a man. Its grand national structures that seem built for eternity--its noble institutions, charitable, and learned, and scientific, and artistical--the genius and science and bravery and moral excellence within its countless walls--have overwhelmed me with a sense of its glory and majesty and power. But in a less admiring mood, I have quite reversed the picture. Perhaps the following sonnet may seem to indicate that the writer while composing it, must have worn his colored spectacles.

LONDON, IN THE MORNING.

> The morning wakes, and through the misty air In sickly radiance struggles--like the dream Of sorrow-shrouded hope. O'er Thames' dull stream, Whose sluggish waves a wealthy burden bear From every port and clime, the pallid glare Of early sun-light spreads. The long streets seem Unpeopled still, but soon each path shall teem With hurried feet, and visages of care. And eager throngs shall meet where dusky marts Resound like ocean-caverns, with the din Of toil and strife and agony and sin. Trade's busy Babel! Ah! how many

hearts By lust of gold to thy dim temples brought In happier hours have scorned the prize they sought?

D.L.R.

I now give a pair of sonnets upon the City of Palaces as viewed through somewhat clearer glasses.

VIEW OF CALCUTTA.

Here Passion's restless eye and spirit rude May greet no kindred images of power To fear or wonder ministrant. No tower, Time-struck and tenantless, here seems to brood, In the dread majesty of solitude, O'er human pride departed-- no rocks lower O'er ravenous billows--no vast hollow wood Rings with the lion's thunder--no dark bower The crouching tiger haunts--no gloomy cave Glitters with savage eyes! But all the scene Is calm and cheerful. At the mild command Of Britain's sons, the skilful and the brave, Fair palace-structures decorate the land, And proud ships float on Hooghly's breast serene!

D.L.R.

SONNET, ON RETURNING TO CALCUTTA AFTER A VOYAGE TO THE STRAITS OF MALACCA.

Umbrageous woods, green dells, and mountains high, And bright cascades, and wide cerulean seas, Slumbering, or snow-wreathed by the freshening breeze, And isles like motionless clouds upon the sky In silent summer noons, late charmed mine eye, Until my soul was stirred like wind-touched trees, And passionate love and speechless ecstasies Up-raised the thoughts in spiritual depths that lie. Fair scenes, ye haunt me still! Yet I behold This sultry city on the level shore Not all unmoved; for here our fathers bold Won proud historic names in days of yore, And here are generous hearts that ne'er grow cold, And many a friendly hand and open door.

D.L.R.

There are several extremely elegant customs connected with some of the Indian Festivals, at which flowers are used in great profusion. The surface of the "sacred river" is often thickly strewn with them. In Mrs. Carshore's pleasing volume of *Songs of the East*[053] there is a long poem (too long to quote entire) in which the *Beara Festival* is described. I must give the introductory passage.

"THE BEARA FESTIVAL.

> "Upon the Ganges' overflowing banks, Where palm trees lined the shore in graceful ranks, I stood one night amidst a merry throng Of British youths and maidens, to behold A witching Indian scene of light and song, Crowds of veiled native loveliness untold, Each streaming path poured duskily along. The air was filled with the sweet breath of flowers, And music that awoke the silent hours, It was the BEARA FESTIVAL and feast When proud and lowly, loftiest and least, Matron and Moslem maiden pay their vows, With impetratory and votive gift, And to the Moslem Jonas bent their brows. *Each brought her floating lamp of flowers*, and swift A thousand lights along the current drift, Till the vast bosom of the swollen stream, Glittering and gliding onward like a dream, Seems a wide mirror of the starry sphere Or more as if the stars had dropt from air, And in an earthly heaven were shining here, And far above were, but reflected there Still group on group, advancing to the brink, As group on group retired link by link; For one pale lamp that floated out of view Five brighter ones they quickly placed anew; At length the slackening multitudes grew less, And the lamps floated scattered and apart. As stars grow few when morning's footsteps press When a slight girl, shy as the timid halt, Not far from where we stood, her offering brought. Singing a low sweet strain, with lips untaught. Her song proclaimed, that 'twas not many hours Since she had left her childhood's innocent home; And now with Beara lamp, and wreathed flowers, To propitiate heaven, for wedded bliss had come"

To these lines Mrs. Carshore (who has been in this country, I believe, from her birth, and who ought to know something of Indian customs) appends the following notes.

"*It was the Beara festival.*" Much has been said about the Beara or floating lamp, but I have never yet seen a correct description. Moore mentions that Lalla Rookh saw a solitary Hindoo girl bring her lamp to the river. D.L.R. says the same, whereas the Beara festival is a Moslem feast that takes place once a year in the monsoons, when thousands of females offer their vows to the patron of rivers.

"*Moslem Jonas*" Khauj Khoddir is the Jonas of the Mussulman; he, like the prophet of Nineveh, was for three days inside a fish, and for that reason is called the patron of rivers."

I suppose Mrs. Carshore alludes, in the first of these notes, to the following passage in the prose part of Lalla Rookh:--

"As they passed along a sequestered river after sunset, they saw a young Hindoo girl upon the bank whose employment seemed to them so strange that they stopped their palanquins to observe her. She had lighted a small lamp, filled with oil of cocoa, and placing it in an earthern dish, adorned with a wreath of flowers, had committed it with a trembling hand to the stream: and was now anxiously watching its progress down the current, heedless of the gay cavalcade which had drawn up beside her. Lalla Rookh was all curiosity;--when one of her attendants, who had lived upon the banks of the Ganges, (where this ceremony is so frequent that often, in the dusk of evening, the river is seen glittering all over with lights, like the Oton-Jala or Sea of Stars,) informed the Princess that it was the usual way, in which the friends of those who had gone on dangerous voyages offered up vows for their safe return. If the lamp sunk immediately, the omen was disastrous; but if it went shining down the stream, and continued to burn till entirely out of sight, the return of the beloved object was considered as certain.

Lalla Rookh, as they moved on, more than once looked back, to observe how the young Hindoo's lamp proceeded: and while she saw with pleasure that it was unextinguished, she could not help fearing that all the hopes of this life were no better than that feeble light upon the river."

Moore prepared himself for the writing of Lalla Rookh by "long and laborious reading." He himself narrates that Sir James Mackintosh was asked by Colonel Wilks, the Historian of British India, whether it was true that the poet had never been in the East. Sir James replied, "*Never.*" "Well, that shows me," said Colonel Wilks, "that reading over D'Herbelot is as good as riding on the back of a camel." Sir John Malcolm, Sir William Ouseley and other high authorities have testified to the accuracy of Moore's descriptions of Eastern scenes and customs.

The following lines were composed on the banks of the Hooghly at Cossipore, (many long years ago) just after beholding the river one evening almost covered with floating lamps.[054]

A HINDU FESTIVAL.

> Seated on a bank of green, Gazing on an Indian scene, I have dreams the mind to cheer, And a feast for eye and ear. At my feet a river flows, And its broad face richly glows With the glory of the sun, Whose proud race is nearly run Ne'er before did sea or stream Kindle thus beneath his beam, Ne'er did miser's eye behold Such a glittering mass of gold 'Gainst the gorgeous radiance float Darkly, many a sloop and boat, While in each the figures seem Like the shadows of a dream Swiftly, passively, they glide As sliders on a frozen tide. Sinks the sun--the sudden night Falls, yet still the scene is bright Now the fire-fly's living spark Glances through the foliage dark, And along the dusky stream Myriad lamps with ruddy gleam On the small waves float and quiver, As if upon the favored river, And to mark the sacred hour, Stars had fallen in a shower. For many a mile is either shore Illumined with a countless store Of lustres ranged in glittering rows, Each a golden column throws To light the dim depths of the tide, And the moon in all her pride Though beauteously her regions glow, Views a scene as fair below

D.L.R.

Mrs. Carshore alludes, I suppose to the above lines, or the following sonnet, or both perhaps, when she speaks of my erroneous Orientalism--

SCENE ON THE GANGES.

> The shades of evening veil the lofty spires Of proud Benares' fanes! A thickening haze Hangs o'er the stream. The weary boatmen raise Along the dusky shore their crimson fires That tinge the circling groups. Now hope inspires Yon Hindu maid, whose heart true passion sways, To launch on Gungas flood the glimmering rays Of Love's frail lamp,--but, lo the light expires! Alas! what sudden sorrow fills her breast! No charm of life remains. Her tears deplore A lover lost and never, never more Shall hope's sweet vision yield her spirit rest! The cold wave quenched the flame--an omen dread That telleth of the faithless--*or the dead*!

D.L.R.

Horace Hayman Wilson, a high authority on all Oriental customs, clearly alludes in the following lines to the launching of floating lamps by *Hindu* females.

> Grave in the tide the Brahmin stands, And folds his cord or twists his hands, And tells his beads, and all unheard Mutters a solemn mystic word With reverence the Sudra dips, And fervently the current sips, That to his humbler hope conveys A future life of happier days. But chief do India's simple daughters Assemble in these hallowed waters, With vase of classic model laden Like Grecian girl or Tuscan maiden, Collecting thus their urns to fill From gushing fount or trickling rill, And still with pious fervour they To Gunga veneration pay And with pretenceless rite prefer, The wishes of their hearts to her The maid or matron, as she throws *Champae* or lotus, *Bel* or rose, Or sends the quivering light afloat In shallow cup or paper boat, Prays for a parent's peace and wealth Prays for a child's success and health, For a fond husband breathes a prayer, For progeny their loves to share, For

what of good on earth is given To lowly life, or hoped in heaven,

H.H.W.

On seeing Miss Carshore's criticism I referred the subject to an intelligent Hindu friend from whom I received the following answer:--

> My dear Sir, The *Beara*, strictly speaking, is a Mahomedan festival. Some of the lower orders of the Hindus of the NW Provinces, who have borrowed many of their customs from the Mahomedans, celebrate the *Beara*. But it is not observed by the Hindus of Bengal, who have a festival of their own, similar to the *Beara*. It takes place on the evening of the *Saraswati Poojah*, when a small piece of the bark of the Plantain Tree is fitted out with all the necessary accompaniments of a boat, and is launched in a private tank with a lamp. The custom is confined to the women who follow it in their own house or in the same neighbourhood. It is called the *Sooa Dooa Breta*. Yours truly,

Mrs. Carshore it would seem is partly right and partly wrong. She is right in calling the *Beara* a *Moslem* Festival. It is so; but we have the testimony of Horace Hayman Wilson to the fact that *Hindu maids and matrons also launch their lamps upon the river*. My Hindu friend acknowledges that his countrymen in the North West Provinces have borrowed many of their customs from the Mahomedans, and though he is not aware of it, it may yet be the case, that some of the Hindus of *Bengal*, as elsewhere, have done the same, and that they set lamps afloat upon the stream to discover by their continued burning or sudden extinction the fate of some absent friend or lover. I find very few Natives who are able to give me any exact and positive information concerning their own national customs. In their explanations of such matters they differ in the most extraordinary manner amongst themselves. Two most respectable and intelligent Native gentlemen who were proposing to lay out their grounds under my directions, told me that I must not cut down a single cocoa-nut tree, as it would be dreadful sacrilege-- equal to cutting the throats of seven brahmins! Another equally respectable and intelligent Native friend, when I mentioned the fact, threw himself back in his chair to give vent to a hearty laugh. When he had recovered himself a little from this risible convulsion he observed that his father and his grandfather had cut down cocoa-nut trees in considerable numbers without the slightest remorse or fear. And yet again, I afterwards heard that one of the richest Hindu families in Calcutta, rather than suffer so sacred an object to be injured, piously submit to a very serious inconvenience occasioned by a cocoa-nut tree standing in the centre of the carriage road that leads to the portico of their large town palace. I am told that there are other sacred trees which must not be removed by the hands of Hindus of inferior caste, though in this case there is a way of getting over the difficulty, for it is allowable or even meritorious to make presents of these trees to Brahmins, who cut them down for their own fire-wood. But the cocoa-nut tree is said to be too sacred even for the axe of a Brahmin.

I have been running away again from my subject;--I was discoursing upon May-day in England. The season there is still a lovely and a merry one, though the most picturesque and romantic of its ancient observances, now live but in the memory of the "oldest inhabitants," or on the page of history.[055]

> See where, amidst the sun and showers, The Lady of the vernal hours, Sweet May, comes forth again with all her flowers.

Barry Cornwall.

The *May-pole* on these days is rarely seen to rise up in English towns with its proper floral decorations[056]. In remote rural districts a solitary May-pole is still, however, occasionally discovered. "A May- pole," says Washington Irving, "gave a glow to my feelings and spread a charm over the country for the rest of the day: and as I traversed a part of the fair plains of Cheshire, and the beautiful borders of Wales and looked from among swelling hills down a long green valley, through which the Deva wound its wizard stream, my imagination turned all into a perfect Arcadia. One can readily imagine what a gay scene old London must have been when the doors were decked with hawthorn; and Robin Hood, Friar Tuck,

Maid Marian, Morris dancers, and all the other fantastic dancers and revellers were performing their antics about the May-pole in every part of the city. I value every custom which tends to infuse poetical feeling into the common people, and to sweeten and soften the rudeness of rustic manners without destroying their simplicity."

Another American writer--a poet--has expressed his due appreciation of the pleasures of the season. He thus addresses the merrie month of MAY.[057]

MAY.

> Would that thou couldst laugh for aye, Merry, ever merry May! Made of sun gleams, shade and showers Bursting buds, and breathing flowers, Dripping locked, and rosy vested, Violet slippered, rainbow crested; Girdled with the eglantine, Festooned with the dewy vine Merry, ever Merry May, Would that thou could laugh for aye!

W.D. Gallagher.

I must give a dainty bit of description from the poet of the poets-- our own romantic Spenser.

> Then comes fair May, the fayrest mayde on ground, Decked with all dainties of the season's pryde, And throwing flowres out of her lap around. Upon two brethren's shoulders she did ride, The twins of Leda, which, on eyther side, Supported her like to their Sovereign queene Lord! how all creatures laught when her they spide, And leapt and danced as they had ravisht beene! And Cupid's self about her fluttred all in greene.

Here are a few lines from Herrick.

> Fled are the frosts, and now the fields appear Re-clothed in freshe and verdant diaper; Thawed are the snowes, and now the lusty spring Gives to each mead a neat enameling, The palmes[058] put forth their gemmes, and every tree Now swaggers in her leavy gallantry.

The Queen of May--Lady Flora--was the British representative of the Heathen Goddess Flora. May still returns and ever will return at her proper season, with all her bright leaves and fragrant blossoms, but men cease to make the same use of them as of yore. England is waxing utilitarian and prosaic.

The poets, let others neglect her as they will, must ever do fitting observance, in songs as lovely and fresh as the flowers of the hawthorn,

> To the lady of the vernal hours.

Poor Keats, who was passionately fond of flowers, and everything beautiful or romantic or picturesque, complains, with a true poet's earnestness, that in *his* day in England there were

> No crowds of nymphs, soft-voiced and young and gay In woven baskets, bringing ears of corn, Roses and pinks and violets, to adorn The shrine of Flora in her early May.

The Floral Games--*Jeux Floraux*--of Toulouse--first celebrated at the commencement of the fourteenth century, are still kept up annually with great pomp and spirit. Clemence Isaure, a French lady, bequeathed to the Academy of Toulouse a large sum of money for the annual celebration of these games. A sort of College Council is formed, which not only confers degrees on those poets who do most honor to the Goddess Flora, but sometimes grants them more substantial favors. In 1324 the poets were encouraged to compete for a golden violet and a silver eglantine and pansy. A century later the prizes offered were an amaranthus of gold of the value of 400 livres, for the best ode, a violet of silver, valued at 250 livres, for an essay in prose, a silver pansy, worth 200 livres, for an eclogue, elegy or idyl, and a silver lily of the value of sixty livres, for the best sonnet or hymn in honor of the Virgin Mary,--for religion is mixed up with merriment, and heathen with Christian rites. He who gained a prize three times was honored with the title of Doctor *en gaye science*, the name given to the poetry of the Provençal troubadours. A mass, a sermon, and alms-giving, commence the ceremonies. The French poet, Ronsard who had gained a prize in the floral games, so delighted Mary Queen of Scots with his verses on the Rose that she presented him with a silver rose worth £500, with this inscription--"*A Ronsard, l'Apollon de la source des Muses.*"

At Ghent floral festivals are held twice a year when amateur and professional florists assemble together and contribute each his share of flowers to the grand general exhibition which is under the direct patronage of the public authorities. Honorary medals are awarded to the possessors of the finest flowers.

The chief floral festival of the Chinese is on their new year's day, when their rivers are covered with boats laden with flowers, and gay flags streaming from every mast. Their homes and temples are richly hung with festoons of flowers. Boughs of the peach and plum trees in blossom, enkíanthus quinque-flòra, camelias, cockscombs, magnolias, jonquils are then exposed for sale in all the streets of Canton. Even the Chinese ladies, who are visible at no other season, are seen on this occasion in flower-boats on the river or in the public gardens on the shore.

The Italians, it is said, still have artificers called *Festaroli*, whose business it is to prepare festoons and garlands. The ancient Romans were very tasteful in their nosegays and chaplets. Pliny tells us that the Sicyonians were especially celebrated for the graceful art exhibited in the arrangement of the varied colors of their garlands, and he gives us the story of Glycera who, to please her lover Pausias, the painter of Sicyon, used to send him the most exquisite chaplets of her own braiding, which he regularly copied on his canvas. He became very eminent as a flower-painter. The last work of his pencil, and his master-piece, was a picture of his mistress in the act of arranging a chaplet. The picture was called the *Garland Twiner*. It is related that Antony for some time mistrusting Cleopatra made her taste in the first instance every thing presented to him at her banquets. One day "the Serpent of old Nile" after dipping her own coronet of flowers into her goblet drank up the wine and then directed him to follow her example. He was off his guard. He dipped his chaplet in his cup. The leaves had been touched with poison. He was just raising the cup to his lips when she seized his arm, and said "Cease your jealous doubts, for know, that if I had desired your death or wished to live without you, I could easily have destroyed you." The Queen then ordered a prisoner to be brought into their presence, who being made to drink from the cup, instantly expired.[059]

Some of the nosegays made up by "flower-girls" in London and its neighbourhood are sold at such extravagant prices that none but the very wealthy are in the habit of purchasing them, though sometimes a poor lover is tempted to present his mistress on a ball-night with a bouquet that he can purchase only at the cost of a good many more leaves of bread or substantial meals than he can well spare. He has to make every day a banian-day for perhaps half a month that his mistress may wear a nosegay for a few hours. However, a lover is often like a cameleon and can almost live on air--*for a time*-- "promise-crammed." 'You cannot feed capons so.'

At Covent Garden Market, (in London) and the first-rate Flower-shops, a single wreath or nosegay is often made up for the head or hand at a price that would support a poor labourer and his family for a month. The colors of the wreaths are artfully arranged, so as to suit different complexions, and so also as to exhibit the most rare and costly flowers to the greatest possible advantage.

All true poets

> --The sages Who have left streaks of light athwart their pages--

have contemplated flowers--with a passionate love, an ardent admiration; none more so than the sweet-souled Shakespeare. They are regarded by the imaginative as the fairies of the vegetable world--the physical personifications of etherial beauty. In *The Winter's Tale* our great dramatic bard has some delightful floral allusions that cannot be too often quoted.

> Here's flowers for you, Hot lavender, mint, savory, majoram, The marigold, that goes to bed with the sun, And with him rises weeping these are flowers Of middle summer, and I think they are given To men of middle age.

> O, Proserpina, For the flowers now that, frighted, thou lett'st fall From Dis's waggon! Daffodils, That come before the swallow dares, and take The winds of March with beauty, violets dim, But sweeter than the lids of Juno's eyes, Or Cytherea's breath, pale primroses, That die unmarried ere they can behold Great Phoebus in his strength,--a malady Most incident to maids, bold oxlips and The crown imperial, lilies of all kinds, The flower de luce being one

Shakespeare here, as elsewhere, speaks of "*pale* primroses." The poets almost always allude to the primrose as a *pale* and interesting invalid. Milton tells us of

> The yellow cowslip and the *pale* primrose[060]

The poet in the manuscript of his *Lycidas* had at first made the primrose "*die unwedded*," which was a pretty close copy of Shakespeare. Milton afterwards struck out the word "*unwedded*," and substituted the word "*forsaken*." The reason why the primrose was said to "die unmarried," is, according to Warton, because it grows in the shade uncherished or unseen by the sun, who was supposed to be in love with certain sorts of flowers. Ben Jonson, however, describes the primrose as *a wedded lady*--"the Spring's own *Spouse*"--though she is certainly more commonly regarded as the daughter of Spring not the wife. J Fletcher gives her the true parentage:--

> Primrose, first born child of Ver

There are some kinds of primroses, that are not *pale*. There is a species in Scotland, which is of a deep purple. And even in England (in some of the northern counties) there is a primrose, the bird's-eye primrose, (Primula farinosa,) of which the blossom is lilac colored and the leaves musk-scented.

In Sweden they call the Primrose *The key of May*.

The primrose is always a great favorite with imaginative and sensitive observers, but there are too many people who look upon the beautiful with a utilitarian eye, or like Wordsworth's Peter Bell regard it with perfect indifference.

A primrose by the river's brim A yellow primrose was to him. And it was nothing more.

I have already given one anecdote of a utilitarian; but I may as well give two more anecdotes of a similar character. Mrs. Wordsworth was in a grove, listening to the cooing of the stock-doves, and associating their music with the remembrance of her husband's verses to a stock-dove, when a farmer's wife passing by exclaimed, "Oh, I do like stock-doves!" The woman won the heart of the poet's wife at once; but she did not long retain it. "Some people," continued the speaker, "like 'em in a pie; for my part I think there's nothing like 'em stewed in inions." This was a rustic utilitarian. Here is an instance of a very different sort of utilitarianism--the utilitarianism of men who lead a gay town life. Sir W.H. listened, patiently for some time to a poetical-minded friend who was rapturously expatiating upon the delicious perfume of a bed of violets; "Oh yes," said Sir W. at last, "its all very well, but for my part I very much prefer the smell of a flambeau at the theatre." But intellects far more capacious than that of Sir W.H. have exhibited the same indifference to the beautiful in nature. Locke and Jeremy Bentham and even Sir Isaac Newton despised all poetry. And yet God never meant man to be insensible to the beautiful or the poetical. "Poetry, like truth," says Ebenezer Elliot, "is a common flower: God has sown it over the earth, like the daisies sprinkled with tears or glowing in the sun, even as he places the crocus and the March frosts together and beautifully mingles life and death." If the finer and more spiritual faculties of men were as well cultivated or exercised as are their colder and coarser faculties there would be fewer utilitarians. But the highest part of our nature is too much neglected in all our systems of education. Of the beauty and fragrance of flowers all earthly creatures except man are apparently meant to be unconscious. The cattle tread down or masticate the fairest flowers without a single "compunctious visiting of nature." This excites no surprize. It is no more than natural. But it is truly painful and humiliating to see any human being as insensible as the beasts of the field to that poetry of the world which God seems to have addressed exclusively to the heart and soul of man.

In South Wales the custom of strewing all kinds of flowers over the graves of departed friends, is preserved to the present day. Shakespeare, it appears, knew something of the customs of that part of his native country and puts the following *flowery* speech into the mouth of the young Prince, Arviragus, who was educated there.

> With fairest flowers, While summer lasts, and I live here, Fidele, I'll sweeten thy sad grave. Thou shalt not lack The flower that's like thy face, pale Primrose, nor The azured Harebell, like thy veins; no, nor The leaf of Eglantine; whom not to slander, Out-sweetened not thy breath.

Cymbeline.

Here are two more flower-passages from Shakespeare.

> Here's a few flowers; but about midnight more; The herbs that have on them cold dew o' the night Are strewings fitt'st for graves.--Upon their faces:-- You were as flowers; now withered; even so These herblets shall, which we upon you strow.

Cymbeline.

> Sweets to the sweet. Farewell! I hoped thou should'st have been my Hamlet's wife; I thought thy bride-bed to have decked, sweet maid, And not t' have strewed thy grave.

Hamlet.

Flowers are peculiarly suitable ornaments for the grave, for as Evelyn truly says, "they are just emblems of the life of man, which has been compared in Holy Scripture to those fading creatures, whose roots being buried in dishonor rise again in glory."[061]

This thought is natural and just. It is indeed a most impressive sight, a most instructive pleasure, to behold some "bright consummate flower" rise up like a radiant exhalation or a beautiful vision-- like good from evil--with such stainless purity and such dainty loveliness, from the hot-bed of corruption.

Milton turns his acquaintance with flowers to divine account in his Lycidas.

> Return; Sicilian Muse, And call the vales, and bid them hither cast Their bells and flowerets of a thousand hues. Ye vallies low, where the mild whispers use Of shades and wanton winds, and gushing brooks, On whose fresh lap the swart-star sparely looks; Throw hither all your quaint enamelled eyes, That on the green turf suck the honied showers. And purple all the ground with vernal flowers. Bring the rathe primrose that forsaken dies. The tufted crow-toe, and pale jessamine, The white pink, and the pansy freaked with jet, The glowing violet, The musk-rose and the well-attired woodbine, With cowslips wan that hang the pensive head,[062] And every flower that sad embroidery wears; Bid Amaranthus all his beauty shed, And daffodillies fill their cups with tears, To strew the laureate hearse where Lycid lies, For, so to interpose a little ease, Let our frail thoughts dally with faint surmise

Here is a nosegay of spring-flowers from the hand of Thomson:--

> Fair handed Spring unbosoms every grace, Throws out the snow drop and the crocus first, the daisy, primrose, violet darkly blue, And polyanthus of unnumbered dyes, The yellow wall flower, stained with iron brown, And lavish stock that scents the garden round, From the soft wing of vernal breezes shed, Anemonies, auriculas, enriched With shining meal o'er all their velvet leaves And full ranunculus of glowing red Then comes the tulip race, where Beauty plays Her idle freaks from family diffused To family, as flies the father dust, The varied colors run, and while they break On the charmed eye, the exulting Florist marks With secret pride, the wonders of his hand Nor gradual bloom is wanting, from the bird, First born of spring, to Summer's musky tribes Nor hyacinth, of purest virgin white, Low bent, and, blushing inward, nor jonquils, Of potent fragrance, nor Narcissus fair, As o'er the fabled fountain hanging still, Nor broad carnations, nor gay spotted pinks; Nor, showered from every bush, the damask rose. Infinite varieties, delicacies, smells, With

hues on hues expression cannot paint, The breath of Nature and her endless bloom.

Here are two bouquets of flowers from the garden of Cowper

> Laburnum, rich In streaming gold, syringa, ivory pure, The scentless and the scented rose, this red, And of an humbler growth, the other[063] tall, And throwing up into the darkest gloom Of neighboring cypress, or more sable yew, Her silver globes, light as the foamy surf That the wind severs from the broken wave, The lilac, various in array, now white, Now sanguine, and her beauteous head now set With purple spikes pyramidal, as if Studious of ornament yet unresolved Which hue she most approved, she chose them all, Copious of flowers the woodbine, pale and wan, But well compensating her sickly looks With never cloying odours, early and late, Hypericum all bloom, so thick a swarm Of flowers, like flies clothing her slender rods, That scarce a loaf appears, mezereon too, Though leafless, well attired, and thick beset With blushing wreaths, investing every spray, Althaea with the purple eye, the broom Yellow and bright, as bullion unalloy'd, Her blossoms, and luxuriant above all The jasmine, throwing wide her elegant sweets, The deep dark green of whose unvarnish'd leaf Makes more conspicuous, and illumines more, The bright profusion of her scatter'd stars

> Th' amomum there[064] with intermingling flowers And cherries hangs her twigs. Geranium boasts Her crimson honors, and the spangled beau Ficoides, glitters bright the winter long All plants, of every leaf, that can endure The winter's frown, if screened from his shrewd bite, Live their and prosper. Those Ausonia claims, Levantine regions those, the Azores send Their jessamine, her jessamine remote Caffraia, foreigners from many lands, They form one social shade as if convened By magic summons of the Orphean lyre

Here is a bunch of flowers laid before the public eye by Mr. Proctor--

There the rose unveils Her breast of beauty, and each delicate bud O' the season comes in turn to bloom and perish, But first of all the violet, with an eye Blue as the midnight heavens, the frail snowdrop, Born of the breath of winter, and on his brow Fixed like a full and solitary star The languid hyacinth, and wild primrose And daisy trodden down like modesty The fox glove, in whose drooping bells the bee Makes her sweet music, the Narcissus (named From him who died for love) the tangled woodbine, Lilacs, and flowering vines, and scented thorns, And some from whom the voluptuous winds of June Catch their perfumings

Barry Cornwall

I take a second supply of flowers from the same hand

> Here, this rose (This one half blown) shall be my Maia's portion, For that like it her blush is beautiful And this deep violet, almost as blue As Pallas' eye, or thine, Lycemnia, I'll give to thee for like thyself it wears Its sweetness, never obtruding. For this lily Where can it hang but it Cyane's breast? And yet twill wither on so white a bed, If flowers have sense of envy.--It shall be Amongst thy raven tresses, Cytheris, Like one star on the bosom of the night The cowslip and the yellow primrose,--they Are gone, my sad Leontia, to their graves, And April hath wept o'er them, and the voice Of March hath sung, even before their deaths The dirge of those young children of the year But here is hearts ease for your woes. And now, The honey suckle flower I give to thee, And love it for my sake, my own Cyane It hangs upon the stem it loves, as thou Hast clung to me, through every joy and sorrow, It flourishes with its guardian growth, as thou dost, And if the woodman's axe should droop the tree, The woodbine too must perish.

Barry Cornwall

Let me add to the above heap of floral beauty a basket of flowers from Leigh Hunt.

Then the flowers on all their beds-- How the sparklers glance their heads, Daisies with their pinky lashes And the marigolds broad flashes, Hyacinth with sapphire bell Curling backward, and the swell Of the rose, full lipped and warm, Bound about whose riper form Her slender virgin train are seen In their close fit caps of green, Lilacs then, and daffodillies, And the nice leaved lesser lilies Shading, like detected light, Their little green-tipt lamps of white; Blissful poppy, odorous pea, With its wing up lightsomely; Balsam with his shaft of amber, Mignionette for lady's chamber, And genteel geranium, With a leaf for all that come; And the tulip tricked out finest, And the pink of smell divinest; And as proud as all of them Bound in one, the garden's gem Hearts-ease, like a gallant bold In his cloth of purple and gold.

Lady Mary Wortley Montague, who introduced inoculation into England--a practically useful boon to us,--had also the honor to be amongst the first to bring from the East to the West an elegant amusement--the Language of Flowers.[065]

> Then he took up his garland, and did show What every flower, as country people hold, Did signify; and how all, ordered thus, Expressed his grief: and, to my thoughts, did read The prettiest lecture of his country art That could be wished.

Beaumont's and Fletcher's "Philaster."

> There from richer banks Culling out flowers, which in a learned order Do become characters, whence they disclose Their mutual meanings, garlands then and nosegays Being framed into epistles.

Cartwright's "Love's Covenant."

> An exquisite invention this, Worthy of Love's most honied kiss, This art of writing *billet-doux* In buds and odours and bright hues, In saying all one feels and thinks In clever daf-

fodils and pinks, Uttering (as well as silence may,) The sweetest words the sweetest way.

Leigh Hunt.

Yet, no--not words, for they But half can tell love's feeling; Sweet flowers alone can say What passion fears revealing.[066] A once bright rose's withered leaf-- A towering lily broken-- Oh, these may paint a grief No words could e'er have spoken.

Moore.

By all those token flowers that tell What words can ne'er express so well.

Byron.

A mystic language, perfect in each part. Made up of bright hued thoughts and perfumed speeches.

Adams.

If we are to believe Shakespeare it is not human beings only who use a floral language:--

Fairies use flowers for their charactery.

Sir Walter Scott tells us that:--

The myrtle bough bids lovers live--

A sprig of hawthorn has the same meaning as a sprig of myrtle: it gives hope to the lover--the sweet heliotrope tells the depth of his passion,--if he would charge his mistress with levity he presents the larkspur,--and a leaf of nettle speaks her cruelty. Poor Ophelia (in *Hamlet*) gives rosemary for remembrance, and pansies (*pensees*) for thoughts. The laurel indicates victory in war or success with the Muses,

"The meed of mighty conquerors and poets sage."

The ivy wreathes the brows of criticism. The fresh vine-leaf cools the hot forehead of the bacchanal. Bergamot and jessamine imply the fragrance of friendship.

The Olive is the emblem of peace--the Laurel, of glory--the Rue, of grace or purification (Ophelia's *Herb of Grace O'Sundays*)--the Primrose, of the spring of human life--the Bud of the White Rose, of Girlhood,--the full blossom of the Red Rose, of consummate beauty--the Daisy, of innocence,--the Butter-cup, of gold--the Houstania, of content--the Heliotrope, of devotion in love--the Cross of Jerusalem, of devotion in religion--the Forget-me-not, of fidelity--the Myrrh, of gladness--the Yew, of sorrow--the Michaelmas Daisy, of cheerfulness in age--the Chinese Chrysanthemum, of cheerfulness in adversity--the Yellow Carnation, of disdain--the Sweet Violet, of modesty--the white Chrysanthemum, of truth--the Sweet Sultan, of felicity--the Sensitive Plant, of maiden shyness--the Yellow Day Lily, of coquetry--the Snapdragon, of presumption--the Broom, of humility--the Amaryllis, of pride--the Grass, of submission--the Fuschia, of taste--the Verbena, of sensibility--the Nasturtium, of splendour--the Heath, of solitude--the Blue Periwinkle, of early friendship--the Honey-suckle, of the bond of love--the Trumpet Flower, of fame--the Amaranth, of immortality--the Adonis, of sorrowful remembrance,--and the Poppy, of oblivion.

The Witch-hazel indicates a spell,--the Cape Jasmine says *I'm too happy*--the Laurestine, *I die if I am neglected*--the American Cowslip, *You are a divinity*--the Volkamenica Japonica, *May you be happy*--the Rose-colored Chrysanthemum, *I love*,--and the Venus' Car, *Fly with me*.

For the following illustrations of the language of flowers I am indebted to a useful and well conducted little periodical published in London and entitled the *Family Friend*;--the work is a great favorite with the fair sex.

"Of the floral grammar, the first rule to be observed is, that the pronoun *I* or *me* is expressed by inclining the symbol flower to the *left*, and the pronoun *thou* or *thee* by inclining it to the *right*. When, however, it is not a real flower offered, but a representation upon paper, these positions must be reversed, so that the symbol leans to the heart of the person whom it is to signify.

The second rule is, that the opposite of a particular sentiment expressed by a flower presented upright is denoted when the symbol is reversed; thus a rose-bud sent upright, with its thorns and leaves, means, "*I fear, but I hope.*" If the bud is returned upside down, it means, "*You must neither hope nor fear.*" Should the thorns, however, be stripped off, the signification is, "*There is everything to hope;*" but if stript of its leaves, "*There is everything to fear.*" By this it will be seen that the expression of almost all flowers may be varied by a change in their positions, or an alteration of their state or condition. For example, the marigold flower placed in the hand signifies "*trouble of spirits;*" on the heart, "*trouble or love;*" on the bosom, "*weariness.*" The pansy held upright denotes "*heart's ease;*" reversed, it speaks the contrary. When presented upright, it says, "*Think of me;*" and when pendent, "*Forget me.*" So, too, the amaryllis, which is the emblem of pride, may be made to express, "*My pride is humbled,*" or, "*Your pride is checked,*" by holding it downwards, and to the right or left, as the sense requires. Then, again, the wallflower, which is the emblem of fidelity in misfortune, if presented with the stalk upward, would intimate that the person to whom it was turned was unfaithful in the time of trouble.

The third rule has relation to the manner in which certain words may be represented; as, for instance, the articles, by tendrils with single, double, and treble branches, as under--

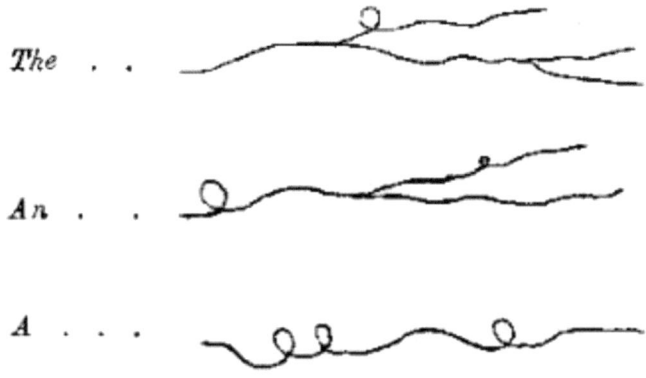

The numbers are represented by leaflets running from one to eleven, as thus--

From eleven to twenty, berries are added to the ten leaves thus--

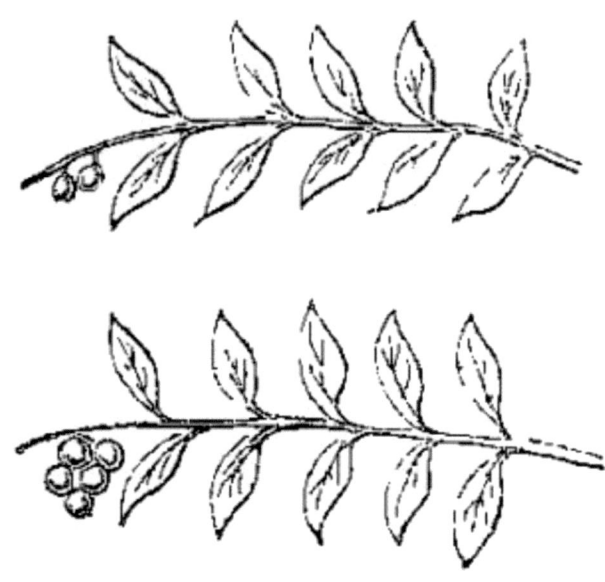

From twenty to one hundred, compound leaves are added to the other ten for the decimals, and berries stand for the odd numbers so--

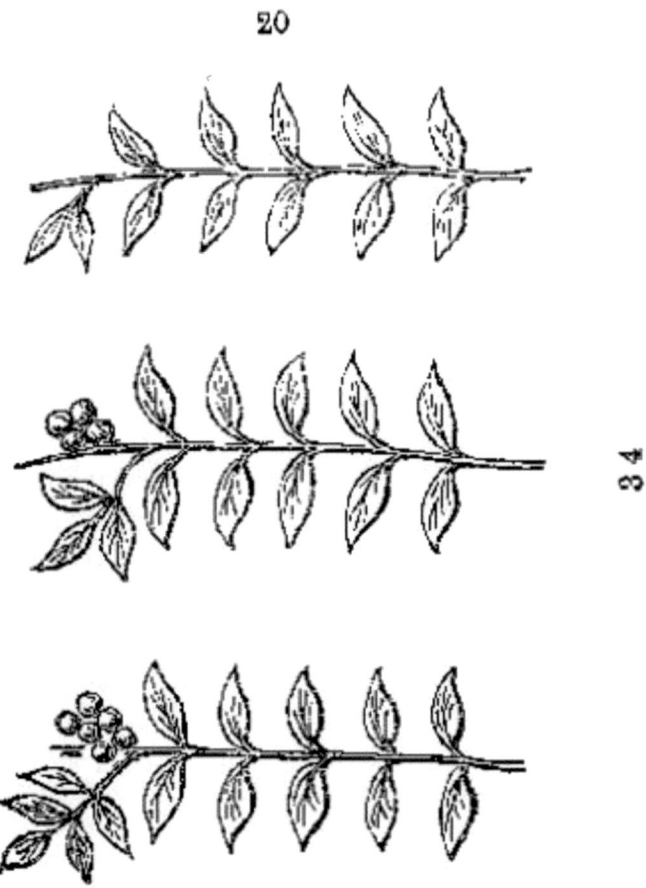

A hundred is represented by ten tens; and this may be increased by a third leaflet and a branch of berries up to 999.

100

A thousand may be symbolized by a frond of fern, having ten or more leaves, and to this a common leaflet may be added to increase the number of thousands. In this way any given number may be represented in foliage, such as the date of a year in which a birthday, or other event, occurs, to which it is desirable to make allusion, in an emblematic wreath or floral picture. Thus, if I presented my love with a mute yet eloquent expression of good wishes on her eighteenth birthday, I should probably do it in this wise:--Within an evergreen wreath (*lasting as my affection*), consisting of ten leaflets and eight berries (*the age of the beloved*), I would place a red rose bud (*pure and lovely*), or a white lily (*pure and modest*), its spotless petals half concealing a ripe strawberry (*perfect excellence*); and to this I might add a blossom of the rose-scented geranium (*expressive of my preference*), a peach blossom to say "*I am your captive*" fern for sincerity, and perhaps bachelor's buttons for *hope in love*"--*Family Friend.*

There are many anecdotes and legends and classical fables to illustrate the history of shrubs and flowers, and as they add something to the peculiar interest with which we regard individual plants, they ought not to be quite passed over by the writers upon Floriculture.

THE FLOS ADONIS.

The Flos Adonis, a blood-red flower of the Anemone tribe, is one of the many plants which, according to ancient story sprang from the tears of Venus and the blood of her coy favorite.

> Rose cheeked Adonis hied him to the chase Hunting he loved, but love he laughed to scorn

Shakespeare.

Venus, the Goddess of Beauty, the mother of Love, the Queen of Laughter, the Mistress of the Graces and the Pleasures, could make no impression on the heart of the beautiful son of Myrrha, (who was changed into a myrrh tree,) though the passion-stricken charmer looked and spake with the lip and eye of the fairest of the immortals. Shakespeare, in his poem of *Venus and Adonis*, has done justice to her burning eloquence, and the lustre of her unequalled loveliness. She had most earnestly, and with all a true lover's care entreated Adonis to avoid the dangers of the chase, but he slighted all her warnings just as he had slighted her affections. He was killed by a wild boar. Shakespeare makes Venus thus lament over the beautiful dead body as it lay on the blood-stained grass.

> Alas, poor world, what treasure hast thou lost! What face remains alive that's worth the viewing? Whose tongue is music now? What can'st thou boast Of things long since, or any thing ensuing? The flowers are sweet, their colors fresh and trim, But true sweet beauty lived and died with him.

In her ecstacy of grief she prophecies that henceforth all sorts of sorrows shall be attendants upon love,--and alas! she was too correct an oracle.

>The course of true love never does run smooth.

Here is Shakespeare's version of the metamorphosis of Adonis into a flower.

> By this the boy that by her side lay killed Was melted into vapour from her sight, And in his blood that on the ground lay spilled, A purple flower sprang up, checquered with white, Resembling well his pale cheeks, and the blood Which in round drops upon their whiteness stood. She bows her head, the new sprung flower to smell, Comparing it to her Adonis' breath, And says, within her bosom it shall dwell Since he himself is reft from her by death; She crops the stalk, and in the branch appears Green dropping sap which she compares to tears.

The reader may like to contrast this account of the change from human into floral beauty with the version of the same story in Ovid as translated by Eusden.

> Then on the blood sweet nectar she bestows, The scented blood in little bubbles rose; Little as rainy drops, which fluttering fly, Borne by the winds, along a lowering sky, Short time ensued, till where the blood was shed, A flower began to rear its purple head Such, as on Punic apples is revealed Or in the filmy rind but half concealed, Still here the fate of lonely forms we see, *So sudden fades the sweet Anemone*. The feeble stems to stormy blasts a prey Their sickly beauties droop, and pine away The winds forbid the flowers to flourish long Which owe to winds their names in Grecian song.

The concluding couplet alludes to the Grecian name of the flower ([Greek: anemos], *anemos*, the wind.)

It is said of the Anemone that it never opens its lips until Zephyr kisses them. Sir William Jones alludes to its short-lived beauty.

> Youth, like a thin anemone, displays His silken leaf, and in a morn decays.

Horace Smith speaks of

> The coy anemone that ne'er discloses Her lips until they're blown on by the wind

Plants open out their leaves to breathe the air just as eagerly as they throw down their roots to suck up the moisture of the earth. Dr. Linley, indeed says, "they feed more by their leaves than their roots." I lately met with a curious illustration of the fact that plants draw a larger proportion of their nourishment from light and air than is commonly supposed. I had a beautiful convolvulus growing upon a trellis work in an upper verandah with a south-western aspect. The root of the plant was in pots. The convolvulus growing too luxuriantly and encroaching too much upon the space devoted to a creeper of another kind, I separated its upper branches from the root and left them to die. The leaves began to fade the second day and most of them were quite dead the third or fourth day, but two or three of the smallest retained a sickly life for some days more. The buds or rather chalices outlived the leaves. The chalices continued to expand every morning, for--I am afraid to say how long a time--it might seem perfectly incredible. The convolvulus is a plant of a rather delicate character and I was perfectly astonished at its tenacity of life in this case. I should mention that this happened in the rainy season and that the upper part of the creeper was partially protected from the sun.

The Anemone seems to have been a great favorite with Mrs. Hemans. She thus addresses it.

> Flower! The laurel still may shed Brightness round the victor's head, And the rose in beauty's hair Still its festal glory wear; And the willow-leaves droop o'er Brows which love sustains no more But by living rays refined, Thou the tremb-

ler of the wind, Thou, the spiritual flower Sentient of each breeze and shower,[067] Thou, rejoicing in the skies And transpierced with all their dyes; Breathing-vase with light o'erflowing, Gem-like to thy centre flowing, Thou the Poet's type shall be Flower of soul, Anemone!

The common anemone was known to the ancients but the finest kind was introduced into France from the East Indies, by Monsieur Bachelier, an eminent Florist. He seems to have been a person of a truly selfish disposition, for he refused to share the possession of his floral treasure with any of his countrymen. For ten years the new anemone from the East was to be seen no where in Europe but in Monsieur Bachelier's parterre. At last a counsellor of the French Parliament disgusted with the florist's selfishness, artfully contrived when visiting the garden to drop his robe upon the flower in such a manner as to sweep off some of the seeds. The servant, who was in his master's secret, caught up the robe and carried it away. The trick succeeded; and the counsellor shared the spoils with all his friends through whose agency the plant was multiplied in all parts of Europe.

THE OLIVE.

The OLIVE is generally regarded as an emblem of peace, and should have none but pleasant associations connected with it, but Ovid alludes to a wild species of this tree into which a rude and licentious fellow was converted as a punishment for "banishing the fair," with indecent words and gestures. The poet tells us of a secluded grotto surrounded by trembling reeds once frequented by the wood-nymphs of the sylvan race:--

> Till Appulus with a dishonest air And gross behaviour, banished thence the fair. The bold buffoon, whene'er they tread the green, Their motion mimics, but with jest obscene; Loose language oft he utters; but ere long A bark in filmy net-work binds his tongue; Thus changed, a base wild olive he remains; The shrub the coarseness of the clown retains.

Garth's Ovid.

The mural of this is excellent. The sentiment reminds me of the Earl of Roscommon's well-known couplet in his *Essay on Translated Verse*, a poem now rarely read.

> Immodest words admit of no defense,[068] For want of decency is want of sense,

THE HYACINTH.

The HYACINTH has always been a great favorite with the poets, ancient and modern. Homer mentions the Hyacinth as forming a portion of the materials of the couch of Jove and Juno.

> Thick new-born Violets a soft carpet spread, And clustering Lotos swelled the rising bed, And sudden *Hyacinths*[069] the turf bestrow, And flaming Crocus made the mountains glow

Iliad, Book 14

Milton gives a similar couch to Adam and Eve.

> Flowers were the couch Pansies, and Violets, and Asphodel And *Hyacinth*, earth's freshest, softest lap

With the exception of the lotus (so common in Hindustan,) all these flowers, thus celebrated by the greatest of Grecian poets, and represented as fit luxuries for the gods, are at the command of the poorest peasant in England. The common Hyacinth is known to the unlearned as the Harebell, so called from the bell shape of its flowers and from its growing so abundantly in thickets frequented by hares. Shakespeare, as we have seen, calls it the *Blue*-bell.

The curling flowers of the Hyacinth, have suggested to our poets the idea of clusters of curling tresses of hair.

> His fair large front and eye sublime declared Absolute rule, and hyacinthine locks Round from his parted forelock manly hung, Clustering

Milton

> The youths whose locks divinely spreading Like vernal hyacinths in sullen hue

Collins

Sir William Jones describes--

> The fragrant hyacinths of Azza's hair, That wanton with the laughing summer air.

A similar allusion may also be found in prose.

"It was the exquisitely fair queen Helen, whose jacinth[070] hair, curled by nature, intercurled by art, like a brook through golden sands, had a rope of fair pearl, which, now hidden by the hair, did, as it were play at fast and loose each with the other, mutually giving and receiving richness."--*Sir Philip Sidney*

"The ringlets so elegantly disposed round the fair countenances of these fair Chiotes [071] are such as Milton describes by 'hyacinthine locks' crisped and curled like the blossoms of that flower"

Dallaway

The old fable about Hyacinthus is soon told. Apollo loved the youth and not only instructed him in literature and the arts, but shared in his pastimes. The divine teacher was one day playing with his pupil at quoits. Some say that Zephyr (Ovid says it was Boreas) jealous of the god's influence over young Hyacinthus, wafted the ponderous iron ring from its right course and caused it to pitch upon the poor boy's head. He fell to the ground a bleeding corpse. Apollo bade the scarlet hyacinth spring from the blood and impressed upon its leaves the words *Ai Ai*, *(alas! alas!)* the Greek funeral lamentation. Milton alludes to the flower in *Lycidas*,

> Like to that sanguine flower inscribed with woe.

Drummond had before spoken of

> That sweet flower that bears In sanguine spots the tenor of our woes

Hurdis speaks of:

> The melancholy Hyacinth, that weeps All night, and never lifts an eye all day.

Ovid, after giving the old fable of Hyacinthus, tells us that "the time shall come when a most valiant hero shall add his name to this flower." "He alludes," says Mr. Riley, "to Ajax, from whose blood when he slew himself, a similar flower[072] was said to have arisen with the letters *Ai Ai* on its leaves, expressive either of grief or denoting the first two letters of his name [Greek: Aias]."

> As poets feigned from Ajax's streaming blood Arose, with grief inscribed, a mournful flower.

Young.

Keats has the following allusion to the old story of Hyacinthus,

> Or they might watch the quoit-pitchers, intent On either side; pitying the sad death Of Hyacinthus, when the cruel breath Of Zephyr slew him,--Zephyr penitent, Who now, ere Phoebus mounts the firmament Fondles the flower amid the sobbing rain.

Endymion.

Our English Hyacinth, it is said, is not entitled to its legendary honors. The words *Non Scriptus* were applied to this plant by Dodonaeus, because it had not the *Ai Ai* upon its petals. Professor Martyn says that the flower called *Lilium Martagon* or the *Scarlet Turk's Cap* is the plant alluded to by the ancients.

Alphonse Karr, the eloquent French writer, whose "*Tour Round my Garden*" I recommend to the perusal of all who can sympathize with reflections and emotions suggested by natural objects, has the following interesting anecdote illustrative of the force of a floral association:--

"I had in a solitary corner of my garden *three hyacinths* which my father had planted and which death did not allow him to see bloom.

Every year the period of their flowering was for me a solemnity, a funeral and religious festival, it was a melancholy remembrance which revived and reblossomed every year and exhaled certain thoughts with its perfume. The roots are dead now and nothing lives of this dear association but in my own heart. But what a dear yet sad privilege man possesses above all created beings, while thus enabled by memory and thought to follow those whom he loved to the tomb and there shut up the living with the dead. What a melancholy privilege, and yet is there one amongst us who would lose it? Who is he who would willingly forget all"

Wordsworth, suddenly stopping before a little bunch of harebells, which along with some parsley fern, grew out of a wall, he exclaimed, 'How perfectly beautiful that is!

> Would that the little flowers that grow could live Conscious of half the pleasure that they give

The Hyacinth has been cultivated with great care and success in Holland, where from two to three hundred pounds have been given for a single bulb. A florist at Haarlem enumerates 800 kinds of double-flowered Hyacinths, besides about 400 varieties of the single kind. It is said that there are altogether upwards of 2000 varieties of the Hyacinth.

The English are particularly fond of the Hyacinth. It is a domestic flower--a sort of parlour pet. When in "close city pent" they transfer the bulbs to glass vases (Hyacinth glasses) filled with water, and place them in their windows in the winter.

An annual solemnity, called Hyacinthia, was held in Laconia in honor of Hyacinthus and Apollo. It lasted three days. So eagerly was this festival honored, that the soldiers of Laconia even when they had taken the field against an enemy would return home to celebrate it.

THE NARCISSUS

> Foolish Narcisse, that likes the watery shore

Spenser

With respect to the NARCISSUS, whose name in the floral vocabulary is the synonyme of *egotism*, there is a story that must be familiar enough to most of my readers. Narcissus was a beautiful youth. Teresias, the Soothsayer, foretold that he should enjoy felicity until he beheld his own face but that the first sight of that would be fatal to him. Every kind of mirror was kept carefully out of his way. Echo was enamoured of him, but he slighted her love, and she pined and withered away until she had nothing left her but her voice, and even that could only repeat the last syllables of other people's sentences. He at last saw his own image reflected in a fountain, and taking it for that of another, he fell passionately in love with it. He attempted to embrace it. On seeing the fruitlessness of all his efforts, he killed himself in despair. When the nymphs raised a funeral pile to burn his body, they found nothing but a flower. That flower (into which he had been changed) still bears his name.

Here is a little passage about the fable, from the *Two Noble Kinsmen* of Beaumont and Fletcher.

> *Emilia*--This garden hath a world of pleasure in it, What flower is this? *Servant*--'Tis called Narcissus, Madam. *Em.*--That was a fair boy certain, but a fool To love himself, were there not maids, Or are they all hard hearted? *Ser*--That could not be to one so fair.

Ben Jonson touches the true moral of the fable very forcibly.

> 'Tis now the known disease That beauty hath, to hear too deep a sense Of her own self conceived excellence Oh! had'st thou known the worth of Heaven's rich gift, Thou would'st have turned it to a truer use, And not (with starved and covetous ignorance) Pined in continual eyeing that bright gem The glance whereof to others had been more Than to thy famished mind the wide world's store.

Gay's version of the fable is as follows:

> Here young Narcissus o'er the fountain stood And viewed his image in the crystal flood The crystal flood reflects his lo-

> vely charms And the pleased image strives to meet his arms. No nymph his inexperienced breast subdued, Echo in vain the flying boy pursued Himself alone, the foolish youth admires And with fond look the smiling shade desires, O'er the smooth lake with fruitless tears he grieves, His spreading fingers shoot in verdant leaves, Through his pale veins green sap now gently flows, And in a short lived flower his beauty glows

Addison has given a full translation of the story of Narcissus from Ovid's Metamorphoses, Book the third.

The common daffodil of our English fields is of the genus Narcissus. "Pray," said some one to Pope, "what is this *Asphodel* of Homer?" "Why, I believe," said Pope "if one was to say the truth, 'twas nothing else but that poor yellow flower that grows about our orchards, and, if so, the verse might be thus translated in English

> --The stern Achilles Stalked through a mead of daffodillies"

THE LAUREL

Daphne was a beautiful nymph beloved by that very amorous gentleman, Apollo. The love was not reciprocal. She endeavored to escape his godship's importunities by flight. Apollo overtook her. She at that instant solicited aid from heaven, and was at once turned into a laurel. Apollo gathered a wreath from the tree and placing it on his own immortal brows, decreed that from that hour the laurel should be sacred to his divinity.

THE SUN-FLOWER

> Who can unpitying see the flowery race Shed by the morn then newflushed bloom resign, Before the parching beam? So fade the fair, When fever revels in their azure veins But one, *the lofty follower of the sun*, Sad when he sits shuts up her yellow leaves, Drooping all night, and when he warm return, Points her enamoured bosom to his ray

Thomson.

THE SUN-FLOWER (*Helianthus*) was once the fair nymph Clytia. Broken-hearted at the falsehood of her lover, Apollo, (who has so many similar sins to answer for) she pined away and died. When it was too late Apollo's heart relented, and in honor of true affection he changed poor Clytia into a *Sun-flower*.[073] It is sometimes called *Tourne-sol*--a word that signifies turning to the sun. Thomas Moore helps to keep the old story in remembrance by the concluding couplet of one of his sweetest ballads.

> Oh! the heart that has truly loved never forgets, But as truly loves on to its close As the sun flower turns on her god when he sets The same look that she turned when he rose

But Moore has here poetized a vulgar error. Most plants naturally turn towards the light, but the sun-flower (in spite of its name) is perhaps less apt to turn itself towards Apollo than the majority of other flowers for it has a stiff stem and a number of heavy heads. At all events it does not change its attitude in the course of the day. The flower-disk that faces the morning sun has it back to it in the evening.

Gerard calls the sun-flower "The Flower of the Sun or the Marigold of Peru". Speaking of it in the year 1596 he tells us that he had some in his own garden in Holborn that had grown to the height of fourteen feet.

THE WALL-FLOWER

> The weed is green, when grey the wall, And blossoms rise where turrets fall

Herrick gives us a pretty version of the story of the WALL-FLOWER, (*cheiranthus cheiri*)("the yellow wall-flower stained with iron brown")

> Why this flower is now called so List sweet maids and you shall know Understand this firstling was Once a brisk and bonny lass Kept as close as Danae was Who a sprightly springal loved, And to have it fully proved, Up she got upon a wall Tempting down to slide withal, But the silken twist

> untied, So she fell, and bruised and died Love in pity of the deed And her loving, luckless speed, Turned her to the plant we call Now, 'The Flower of the Wall'

The wall-flower is the emblem of fidelity in misfortune, because it attaches itself to fallen towers and gives a grace to ruin. David Moir (the Delta of *Blackwood's Magazine*) has a poem on this flower. I must give one stanza of it.

> In the season of the tulip cup When blossoms clothe the trees, How sweet to throw the lattice up And scent thee on the breeze; The butterfly is then abroad, The bee is on the wing, And on the hawthorn by the road The linnets sit and sing.

Lord Bacon observes that wall-flowers are very delightful when set under the parlour window or a lower chamber window. They are delightful, I think, any where.

THE JESSAMINE.

> The Jessamine, with which the Queen of flowers, To charm her god[074] adorns his favorite bowers, Which brides, by the plain hand of neatness dressed-- Unenvied rivals!--wear upon their breast; Sweet as the incense of the morn, and chaste As the pure zone which circles Dian's waist.

Churchill.

The elegant and fragrant JESSAMINE, or Jasmine, (*Jasmimum Officinale*) with its "bright profusion of scattered stars," is said to have passed from East to West. It was originally a native of Hindustan, but it is now to be found in every clime, and is a favorite in all. There are many varieties of it in Europe. In Italy it is woven into bridal wreaths and is used on all festive occasions. There is a proverbial saying there, that she who is worthy of being decorated with jessamine is rich enough for any husband. Its first introduction into that sunny land is thus told. A certain Duke of Tuscany, the first possessor of a plant of this tribe, wished to preserve it as an unique, and forbade his gardener to give away a single sprig of it. But the gardener was a more faithful lover than servant and was more willing to please a young mistress than an old master. He presented the young girl with a branch of jessamine on her birth-day. She planted it in the ground; it took root, and grew and blossomed. She multiplied the plant by cuttings, and by the sale of these realized a little fortune, which her lover received as her marriage dowry.

In England the bride wears a coronet of intermingled orange blossom and jessamine. Orange flowers indicate chastity, and the jessamine, elegance and grace.

THE ROSE.

> For here the rose expands Her paradise of leaves.

Southey.

The ROSE, (*Rosa*) the Queen of Flowers, was given by Cupid to Harpocrates, the God of Silence, as a bribe, to prevent him from betraying the amours of Venus. A rose suspended from the ceiling intimates that all is strictly confidential that passes under it. Hence the phrase--*under the Rose*[075].

The rose was raised by Flora from the remains of a favorite nymph. Venus and the Graces assisted in the transformation of the nymph into a flower. Bacchus supplied streams of nectar to its root, and Vertumnus showered his choicest perfumes on its head.

The loves of the Nightingale and the Rose have been celebrated by the Muses of many lands. An Eastern poet says "You may place a hundred handfuls of fragrant herbs and flowers before the Nightingale; yet he wishes not, in his constant heart, for more than the sweet breath of his beloved Rose."

The Turks say that the rose owes its origin to a drop of perspiration that fell from the person of their prophet Mahommed.

The classical legend runs that the rose was at first of a pure white, but a rose-thorn piercing the foot of Venus when she was hastening to protect Adonis from the rage of Mars, her blood dyed the flower. Spenser alludes to this legend:

> White as the native rose, before the change Which Venus' blood did on her leaves impress.

Spenser.

Milton says that in Paradise were,

> Flowers of all hue, and *without thorns the rose.*

According to Zoroaster there was no thorn on the rose until Ahriman (the Evil One) entered the world.

Here is Dr. Hooker's account of the origin of the red rose.

> To sinless Eve's admiring sight The rose expanded snowy white, When in the ecstacy of bliss She gave the modest flower a kiss, And instantaneous, lo! it drew From her red lip its blushing hue; While from her breath it sweetness found, And spread new fragrance all around.

This reminds me of a passage in Mrs. Barrett Browning's *Drama of Exile* in which she makes Eve say--

> --For was I not At that last sunset seen in Paradise, When all the westering clouds flashed out in throngs Of sudden angel-faces, face by face, All hushed and solemn, as a thought of God Held them suspended,--was I not, that hour The lady of the world, princess of life, Mistress of feast and favour? *Could I touch A Rose with my white hand, but it became Redder at once?*

Another poet. (Mr. C. Cooke) tells us that a species of red rose with all her blushing honors full upon her, taking pity on a very pale maiden, changed complexions with the invalid and became herself as white as snow.

Byron expressed a wish that all woman-kind had but one *rosy* mouth, that he might kiss all woman-kind at once. This, as some one has rightly observed, is better than Caligula's wish that all mankind had but one head that he might cut it off at a single blow.

Leigh Hunt has a pleasant line about the rose:

> And what a red mouth hath the rose, the woman of the flowers!

In the Malay language the same word signifies *flowers* and *women*.

Human beauty and the rose are ever suggesting images of each other to the imagination of the poets. Shakespeare has a beautiful description of the two little princes sleeping together in the Tower of London.

> Their lips were four red roses on a stalk That in their summer beauty kissed each other.

William Browne (our Devonshire Pastoral Poet) has a *rosy* description of a kiss:--

> To her Amyntas Came and saluted; never man before More blest, nor like this kiss hath been another But when two dangling cherries kist each other; Nor ever beauties, like, met at such closes, But in the kisses of two damask roses.

Here is something in the same spirit from Crashaw.

> So have I seen Two silken sister-flowers consult and lay Their bashful cheeks together; newly they Peeped from their buds, showed like the garden's eyes Scarce waked, like was the crimson of their joys, Like were the tears they wept, so like that one Seemed but the other's kind reflection.

Loudon says that there is a rose called the *York and Lancaster* which when, it comes true has one half of the flower red and the other half white. It was named in commemoration of the two houses at the marriage of Henry VII. of Lancaster with Elizabeth of York.

Anacreon devotes one of his longest and best odes to the laudation of the Rose. Such innumerable translations have been made of it that it is now too well known for quotation in this place. Thomas Moore in his version of the ode gives in a foot-note the following translation of a fragment of the Lesbian poetess.

> If Jove would give the leafy bowers A queen for all their world of flowers The Rose would be the choice of Jove, And blush the queen of every grove Sweetest child of weeping

morning, Gem the vest of earth adorning, Eye of gardens, light of lawns, Nursling of soft summer dawns June's own earliest sigh it breathes, Beauty's brow with lustre wreathes, And to young Zephyr's warm caresses Spreads abroad its verdant tresses, Till blushing with the wanton's play Its cheeks wear e'en a redder ray.

From the idea of excellence attached to this Queen of Flowers arose, as Thomas Moore observes, the pretty proverbial expression used by Aristophanes--*you have spoken roses*, a phrase adds the English poet, somewhat similar to the *dire des fleurettes* of the French.

The Festival of the Rose is still kept up in many villages of France and Switzerland. On a certain day of every year the young unmarried women assemble and undergo a solemn trial before competent judges, the most virtuous and industrious girl obtains a crown of roses. In the valley of Engandine, in Switzerland, a man accused of a crime but proved to be not guilty, is publicly presented by a young maiden with a white rose called the Rose of Innocence.

Of the truly elegant Moss Rose I need say nothing myself; it has been so amply honored by far happier pens than mine. Here is a very ingenious and graceful story of its origin. The lines are from the German.

THE MOSS ROSE

The Angel of the Flowers one day, Beneath a rose tree sleeping lay, The spirit to whom charge is given To bathe young buds in dews of heaven, Awaking from his light repose The Angel whispered to the Rose "O fondest object of my care Still fairest found where all is fair, For the sweet shade thou givest to me Ask what thou wilt 'tis granted thee" "Then" said the Rose, "with deepened glow On me another grace bestow." The spirit paused in silent thought What grace was there the flower had not? 'Twas but a moment--o'er the rose A veil of moss the Angel throws, And robed in Nature's simple weed, Could there a flower that rose exceed?

Madame de Genlis tells us that during her first visit to England she saw a moss-rose for the first time in her life, and that when she took it back to Paris it gave great delight to her fellow-citizens, who said it was the first that had ever been seen in that city. Madame de Latour says that Madame de Genlis was mistaken, for the moss-rose came originally from Provence and had been known to the French for ages.

The French are said to have cultivated the Rose with extraordinary care and success. It was the favorite flower of the Empress Josephine, who caused her own name to be traced in the parterres at Malmaison with a plantation of the rarest roses. In the royal rosary at Versailles there are standards eighteen feet high grafted with twenty different varieties of the rose.

With the Romans it was no metaphor but an allusion to a literal fact when they talked of sleeping upon beds of roses. Cicero in his third oration against Verres, when charging the proconsul with luxurious habits, stated that he had made the tour of Sicily seated upon roses. And Seneca says, of course jestingly, that a Sybarite of the name of Smyrndiride was unable to sleep if one of the rose-petals on his bed happened to be curled! At a feast which Cleopatra gave to Marc Antony the floor of the hall was covered with fresh roses to the depth of eighteen inches. At a fête given by Nero at Baiae the sum of four millions of sesterces or about 20,000*l*. was incurred for roses. The Natives of India are fond of the rose, and are lavish in their expenditure at great festivals, but I suppose that no millionaire amongst them ever spent such an amount of money as this upon flowers alone.[076]

I shall close the poetical quotations on the Rose with one of Shakespeare's sonnets.

> O how much more doth beauty beauteous seem, By that sweet ornament which truth doth give. The rose looks fair, but fairer we it deem For that sweet odour which doth in it live. The canker-blooms have full as deep a dye As the perfumed tincture of the roses, Hang on such thorns, and play as wantonly, When summer's breath their masked buds discloses; But for their virtue only is their show, They live unwoo'd

and unrespected fade; Die to themselves. Sweet Roses do not so; Of then sweet deaths are sweetest odours made: And so of you, beauteous and lovely youth, When that shall fade, my verse distils your truth.

There are many hundred acres of rose trees at Ghazeepore which are cultivated for distillation, and making "attar." There are large fields of roses in England also, for the manufacture of rose-water.

There is a story about the origin of attar of Roses. The Princess Nourmahal caused a large tank, on which she used to be rowed about with the great Mogul, to be filled with rose-water. The heat of the sun separating the water from the essential oil of the rose, the latter was observed to be floating on the surface. The discovery was immediately turned to good account. At Ghazeepoor, the *essence*, *atta* or *uttar* or *otto*, or whatever it should be called, is obtained with great simplicity and ease. After the rose water is prepared it is put into large open vessels which are left out at night. Early in the morning the oil that floats upon the surface is skimmed off, or sucked up with fine dry cotton wool, put into bottles, and carefully sealed. Bishop Heber says that to produce one rupee's weight of atta 200,000 well grown roses are required, and that a rupee's weight sells from 80 to 100 rupees. The atta sold in Calcutta is commonly adulterated with the oil of sandal wood.

LINNAEA BOREALIS

The LINNAEA BOREALIS, or two horned Linnaea, though a simple Lapland flower, is interesting to all botanists from its association with the name of the Swedish Sage. It has pretty little bells and is very fragrant. It is a wild, unobtrusive plant and is very averse to the trim lawn and the gay flower-border. This little woodland beauty pines away under too much notice. She prefers neglect, and would rather waste her sweetness on the desert air, than be introduced into the fashionable lists of Florist's flowers. She shrinks from exposure to the sun. A gentleman after walking with Linnaeus on the shores of the lake near Charlottendal on a lovely evening, writes thus "I gathered a small flower and asked if it was the *Linnaea borealis*. 'Nay,' said the philosopher, 'she lives not here, but in the middle of our largest woods. She clings with her little arms to the moss, and seems to resist very gently if you force her from it. She has a complexion like a milkmaid, and ah! she is very, very sweet and agreeable!"

THE FORGET-ME-NOT

The dear little FORGET-ME-NOT, (*myosotis palustris*)[077] with its eye of blue, is said to have derived its touching appellation from a sentimental German story. Two lovers were walking on the bank of a rapid stream. The lady beheld the flower growing on a little island, and expressed a passionate desire to possess it. He gallantly plunged into the stream and obtained the flower, but exhausted by the force of the tide, he had only sufficient strength left as he neared the shore to fling the flower at the fair one's feet, and exclaim "*Forget-me-not!*" (*Vergiss-mein-nicht.*) He was then carried away by the stream, out of her sight for ever.

THE PERIWINKLE.

The PERIWINKLE (*vinca* or *pervinca*) has had its due share of poetical distinction. In France the common people call it the Witch's violet. It seems to have suggested to Wordsworth an idea of the consciousness of flowers.

> Through primrose tufts, in that sweet bower, The Periwinkle trailed its wreaths, *And 'tis my faith that every flower Enjoys the air it breathes.*

Mr. J.L. Merritt, has some complimentary lines on this flower.

> The Periwinkle with its fan-like leaves All nicely levelled, is a lovely flower Whose dark wreath, myrtle like, young Flora weaves; There's none more rare Nor aught more meet to deck a fairy's bower Or grace her hair.

The little blue Periwinkle is rendered especially interesting to the admirers of the genius of Rousseau by an anecdote that records his emotion on meeting it in one of his botanical excursions. He had seen it thirty years before in company with Madame de Warens. On meeting its sweet face again, after so long and eventful an interim, he fell upon his knees, crying out--*Ah! voila de la pervanche!* "It struck him," says Hazlitt, "as the same little identical flower that he remembered so well; and thirty years of sorrow and bitter regret were effaced from his memory."

The Periwinkle was once supposed to be a cure for many diseases. Lord Bacon says that in his time people afflicted with cramp wore bands of green periwinkle tied about their limbs. It had also its supposed moral influences. According to Culpepper the leaves of the flower if eaten by man and wife together would revive between them a lost affection.

THE BASIL.

> Sweet marjoram, with her like, *sweet basil*, rare for smell.

Drayton.

The BASIL is a plant rendered poetical by the genius which has handled it. Boccaccio and Keats have made the name of the *sweet basil* sound pleasantly in the ears of many people who know nothing of botany. A species of this plant (known in Europe under the botanical name of *Ocymum villosum*, and in India as the *Toolsee*) is held sacred by the Hindus. Toolsee was a disciple of Vishnu. Desiring to be his wife she excited the jealousy of Lukshmee by whom she was transformed into the herb named after her.[078]

THE TULIP.

> Tulips, like the ruddy evening streaked.

Southey.

The TULIP (*tulipa*) is the glory of the garden, as far as color without fragrance can confer such distinction. Some suppose it to be 'The Lily of the Field' alluded to in the Sermon on the Mount. It grows wild in Syria.

The name of the tulip is said to be of Turkish origin. It was called Tulipa from its resemblance to the tulipan or turban.

> What crouds the rich Divan to-day With turbaned heads, of every hue Bowing before that veiled and awful face Like Tu-

> lip-beds of different shapes and dyes, Bending beneath the invisible west wind's sighs?

Moore.

The reader has probably heard of the Tulipomania once carried to so great an excess in Holland.

> With all his phlegm, it broke a Dutchman's heart, At a vast price, with one loved root to part.

Crabbe.

About the middle of the 17th century the city of Haarlem realized in three years ten millions sterling by the sale of tulips. A single tulip (the *Semper Augustus*) was sold for one thousand pounds. Twelve acres of land were given for a single root and engagements to the amount of £5,000 were made for a first-class tulip when the mania was at its height. A gentleman, who possessed a tulip of great value, hearing that some one was in possession of a second root of the same kind, eagerly secured it at a most extravagant price. The moment he got possession of it, he crushed it under his foot. "Now," he exclaimed, "my tulip is unique!"

A Dutch Merchant gave a sailor a herring for his breakfast. Jack seeing on the Merchant's counter what he supposed to be a heap of onions, took up a handful of them and ate them with his fish. The supposed onions were tulip bulbs of such value that they would have paid the cost of a thousand Royal feasts.[079]

The tulip mania never leached so extravagant a height in England as in Holland, but our country did not quite escape the contagion, and even so late as the year 1836 at the sale of Mr. Clarke's tulips at Croydon, seventy two pounds were given for a single bulb of the *Fanny Kemble*; and a Florist in Chelsea in the same year, priced a bulb in his catalogue at 200 guineas.

The Tulip is not endeared to us by many poetical associations. We have read, however, one pretty and romantic tale about it. A poor old woman who lived amongst the wild hills of Dartmoor, in Devonshire, possessed a beautiful bed of Tulips, the pride of her small garden. One fine moonlight night her attention was arrested by the sweet music which seemed to issue from a thousand Liliputian choristers. She found that the sounds proceeded from her many colored bells of Tulips. After watching the flowers intently she perceived that they were not swayed to and fro by the wind, but by innumerable little beings that were climbing on the stems and leaves. They were pixies. Each held in its arms an elfin baby tinier than itself. She saw the babies laid in the bells of the plant, which were thus used as cradles, and the music was formed of many lullabies. When the babies were asleep the pixies or fairies left them, and gamboled on the neighbouring sward on which the old lady discovered the day after, several new green rings,--a certain evidence that her fancy had not deceived her! At earliest dawn the fairies had returned to the tulips and taken away their little ones. The good old woman never permitted her tulip bed to be disturbed. She regarded it as holy ground. But when she died, some Utilitarian gardener turned it into a parsley bed! The parsley never flourished. The ground was now cursed. In gratitude to the memory of the benevolent dame who had watched and protected the floral nursery, every month, on the night before the full moon, the fairies scattered flowers on her grave, and raised a sweet musical dirge--heard only by poetic ears--or by maids and children who

> Hold each strange tale devoutly true.

For as the poet says:

> What though no credit doubting wits may give, The fair and innocent shall still believe.

Men of genius are often as trustful as maids and children. Collins, himself a lover of the wonderful, thus speaks of Tasso:--

> Prevailing poet! whose undoubting mind Believed the magic wonders that he sung.

All nature indeed is full of mystery to the imaginative.

> And visions as poetic eyes avow Hang on each leaf and cling to every bough.

The Hindoos believe that the Peepul tree of which the foliage trembles like that of the aspen, has a spirit in every leaf.

"Did you ever see a fairy's funeral, Madam?" said Blake, the artist. "Never Sir." "*I* have," continued that eccentric genius, "One night I was walking alone in my garden. There was great stillness amongst the branches and flowers and more than common sweetness in the air. I heard a low and pleasant sound, and knew not whence it came: at last I perceived *the broad leaf of a flower move*, and underneath I saw a procession of creatures the size and color of green and gray grasshoppers, *bearing a body laid out on a rose leaf*, which they buried with song, and then disappeared."

THE PINK.

The PINK (*dianthus*) is a very elegant flower. I have but a short story about it. The young Duke of Burgundy, grandson of Louis the Fifteenth, was brought up in the midst of flatterers as fulsome as those rebuked by Canute. The youthful prince was fond of cultivating pinks, and one of his courtiers, by substituting a floral changeling, persuaded him that one of those pinks planted by the royal hand had sprung up into bloom in a single night! One night, being unable to sleep, he wished to rise, but was told that it was midnight; he replied *"Well then, I desire it to be morning."*

The pink is one of the commonest of the flowers in English gardens. It is a great favorite all over Europe. The botanists have enumerated about 400 varieties of it.

THE PANSY OR HEARTS-EASE.

The PANSY (*víola trîcolor*) commonly called *Hearts-ease*, or *Love-in- idleness*, or *Herb-Trinity* (*Flos Trinitarium*), or *Three-faces- under-a-hood*, or *Kit-run-about*, is one of the richest and loveliest of flowers.

The late Mrs. Siddons, the great actress, was so fond of this flower that she thought she could never have enough of it. Besides round beds of it she used it as an edging to all the flower borders in her garden. She liked to plant a favorite flower in large masses of beauty. But such beauty must soon fatigue the eye with its sameness. A round bed of one sort of flowers only is like a nosegay composed of one sort of flowers or of flowers of the same hue. She was also particularly fond of evergreens because they gave her garden a pleasant aspect even in the winter.

"Do you hear him?"--(John Bunyan makes the guide enquire of Christiana while a shepherd boy is singing beside his sheep)--"I will dare to say this boy leads a merrier life, and wears more of the herb called *hearts-ease* in his bosom, than he that is clothed in silk and purple."

Shakespeare has connected this flower with a compliment to the maiden Queen of England.

> That very time I saw (but thou couldst not) Flying between the cold moon and the earth, Cupid all armed, a certain aim he took At a fair Vestal, throned by the west; And loosed his love-shaft smartly from his bow As it should pierce a hundred thousand hearts. But I might see young Cupid's fiery shaft Quenched in the chaste beams of the watery moon-- And the imperial votaress passed on In maiden meditation fancy free, Yet marked I where the bolt of Cupid fell. It fell upon *a little western flowers, Before milk white, now purple with love's wound-- And maidens call it* LOVE IN IDLENESS Fetch me that flower, the herb I showed thee once, The juice of it on sleeping eyelids laid, Will make or man or woman madly dote Upon the next live creature that it sees. Fetch me this herb and be thou here again, Ere the leviathan can swim a league.

Midsummer Night's Dream.

The hearts-ease has been cultivated with great care and success by some of the most zealous flower-fanciers amongst our countrymen in India. But it is a delicate plant in this clime, and requires most assiduous attention, and a close study of its habits. It always withers here under ordinary hands.

THE MIGNONETTE.

The MIGNONETTE, (*reseda odorato,*) the Frenchman's *little darling*, was not introduced into England until the middle of the 17th century. The Mignonette or Sweet Reseda was once supposed capable of assuaging pain, and of ridding men of many of the ills that flesh is heir to. It was applied with an incantation. This flower has found a place in the armorial bearings of an illustrious family of Saxony. I must tell the story: The Count of Walsthim loved the fair and sprightly Amelia de Nordbourg. She was a spoilt child and a coquette. She had an humble companion whose christian name was Charlotte. One evening at a party, all the ladies were called upon to choose a flower each, and the gentlemen were to make verses on the selections. Amelia fixed upon the flaunting rose, Charlotte the modest mignonette. In the course of the evening Amelia coquetted so desperately with a dashing Colonel that the Count could not suppress his vexation. On this he wrote a verse for the Rose:

> Elle ne vit qu'un jour, et ne plait qu'un moment. (She lives but for a day and pleases but for a moment)

He then presented the following line on the Mignonette to the gentle Charlotte:

> "Ses qualities surpassent ses charmes."

The Count transferred his affections to Charlotte, and when he married her, added a branch of the Sweet Reseda to the ancient arms of his family, with the motto of

> Your qualities surpass your charms.

VERVAIN.

> The vervain-- That hind'reth witches of their will.

Drayton

VERVAIN (*verbena*) was called by the Greeks *the sacred herb*. It was used to brush their altars. It was supposed to keep off evil spirits. It was also used in the religious ceremonies of the Druids and is still held sacred by the Persian Magi. The latter lay branches of it on the altar of the sun.

The ancients had their *Verbenalia* when the temples were strewed with vervain, and no incantation or lustration was deemed perfect without the aid of this plant. It was supposed to cure the bite of a serpent or a mad dog.

THE DAISY.

The DAISY or day's eye (*bellis perennis*) has been the darling of the British poets from Chaucer to Shelley. It is not, however, the darling of poets only, but of princes and peasants. And it is not man's favorite only, but, as Wordsworth says, Nature's favorite also. Yet it is "the simplest flower that blows." Its seed is broadcast on the land. It is the most familiar of flowers. It sprinkles every field and lane in the country with its little mimic stars. Wordsworth pays it a beautiful compliment in saying that

> Oft alone in nooks remote *We meet it like a pleasant thought When such is wanted.*

But though this poet dearly loved the daisy, in some moods of mind he seems to have loved the little celandine (common pilewort) even better. He has addressed two poems to this humble little flower. One begins with the following stanza.

> Pansies, Lilies, Kingcups, Daisies, Let them live upon their praises; Long as there's a sun that sets Primroses will have their glory; Long as there are Violets, They will have a place in story: There's a flower that shall be mine, 'Tis the little Celandine.

No flower is too lowly for the affections of Wordsworth. Hazlitt says, "the daisy looks up to Wordsworth with sparkling eye as an old acquaintance; a withered thorn is weighed down with a heap of recollections; and even the lichens on the rocks have a life and being in his thoughts."

The Lesser Celandine, is an inodorous plant, but as Wordsworth possessed not the sense of smell, to him a deficiency of fragrance in a flower formed no objection to it. Miss Martineau alludes to a newspaper report that on one occasion the poet suddenly found himself capable of enjoying the fragrance of a flower, and gave way to an emotion of tumultuous rapture. But I have seen this contradicted. Miss Martineau herself has generally no sense of smell, but we have her own testimony to the fact that a brief enjoyment of the faculty once actually occurred to her. In her case there was a simultaneous awakening of two dormant faculties-- the sense of smell and the sense of taste. Once and once only, she enjoyed the scent of a bottle of Eau de Cologne and the taste of meat. The two senses died away again almost in their birth.

Shelley calls Daisies "those pearled Arcturi of the earth"--"the constellated flower that never sets."

The Father of English poets does high honor to this star of the meadow in the "Prologue to the Legend of Goode Women."

He tells us that in the merry month of May he was wont to quit even his beloved books to look upon the fresh morning daisy.

> Of all the floures in the mede Then love I most these floures white and red, Such that men callen Daisies in our town, To them I have so great affectión. As I sayd erst, when comen is the Maie, That in my bedde there dawneth me no daie That I nam up and walking in the mede To see this floure agenst the Sunne sprede, When it up riseth early by the morrow That blisfull sight softeneth all my sorrow.

Chaucer.

The poet then goes on with his hearty laudation of this lilliputian luminary of the fields, and hesitates not to describe it as "of all floures the floure." The famous Scottish Peasant loved it just as truly, and did it equal honor. Who that has once read, can ever forget his harmonious and pathetic address to a mountain daisy on turning it up with the plough? I must give the poem a place here, though it must be familiar to every reader. But we can read it again and again, just as we can look day after day with undiminished interest upon the flower that it commemorates.

Mrs. Stowe (the American writer) observes that "the daisy with its wide plaited ruff and yellow centre is not our (that is, an American's) flower. The English flower is the

> Wee, modest, crimson tippéd flower

which Burns celebrated. It is what we (in America) raise in greenhouses and call the Mountain Daisy. Its effect, growing profusely about fields and grass-plats, is very beautiful."

TO A MOUNTAIN DAISY.

ON TURNING ONE DOWN WITH THE PLOUGH IN APRIL, 1786

> Wee, modest, crimson tippéd flow'r, Thou's met me in an evil hour, For I maun[080] crush amang the stoure[081] Thy slender stem, To spare thee now is past my pow'r, Thou bonnie gem. Alas! its no thy neobor sweet, The bonnie lark, companion meet, Bending thee 'mang the dewy weet[082] Wi' speckled breast, When upward springing, blythe, to greet The purpling east Cauld blew the bitter biting north Upon thy early, humble, birth, Yet cheerfully thou glinted[083] forth Amid the storm, Scarce reared above the patient earth Thy tender form The flaunting flowers our gardens yield, High sheltering woods and wa's[084] maun shield, But thou beneath the random bield[085] O' clod or stane, Adorns the histie[086] stibble field[087] Unseen, alane. There, in thy scanty mantle clad, Thy snawye bosom sun ward spread, Thou lifts thy unassuming head In humble guise, But now the share up tears thy bed, And low thou lies! Such is the fate of artless Maid, Sweet flow'ret of the rural shade! By love's simplicity betrayed, And guileless trust, Till she, like thee, all soiled is laid Low i' the dust. Such is the fate of simple Bard, On Life's rough ocean luckless starred! Unskilful he to note the card Of prudent lore, Till billows rage, and gales blow hard And whelm him o'er! Such fate to suffering worth is given Who long with wants and woes has striven By human pride or cunning driven To misery's brink, Till wrenched of every stay but Heaven, He, ruined, sink! Ev'n thou who mourn'st the Daisy's fate, That fate is thine--no distant date; Stern Ruin's plough-share drives elate, Full on thy bloom; Till crushed beneath the furrow's weight Shall be thy doom.

Burns.

The following verses though they make no pretension to the strength and pathos of the poem by the great Scottish Peasant, have a grace and simplicity of their own, for which they have long been deservedly popular.

A FIELD FLOWER.

ON FINDING ONE IN FULL BLOOM, ON CHRISTMAS DAY, 1803.

> There is a flower, a little flower, With silver crest and golden eye, That welcomes every changing hour, And weathers every sky. The prouder beauties of the field In gay but quick succession shine, Race after race their honours yield, They flourish and decline. But this small flower, to Nature dear, While moons and stars their courses run, Wreathes the whole circle of the year, Companion of the sun. It smiles upon the lap of May, To sultry August spreads its charms, Lights pale October on his way, And twines December's arms. The purple heath and golden broom, On moory mountains catch the gale, O'er lawns the lily sheds perfume, The violet in the vale. But this bold floweret climbs the hill, Hides in the forest, haunts the glen, Plays on the margin of the rill, Peeps round the fox's den. Within the garden's cultured round It shares the sweet carnation's bed; And blooms on consecrated ground In honour of the dead. The lambkin crops its crimson gem, The wild-bee murmurs on its breast, The blue-fly bends its pensile stem, Light o'er the sky-lark's nest. 'Tis FLORA'S page,--in every place, In every season fresh and fair; It opens with perennial grace. And blossoms everywhere. On waste and woodland, rock and plain, Its humble buds unheeded rise; The rose has but a summer-reign; The DAISY never dies.

James Montgomery.

Montgomery has another very pleasing poetical address to the daisy. The poem was suggested by the first plant of the kind which had appeared in India. The flower sprang up unexpectedly out of some English earth, sent with other seeds in it, to this country. The amiable Dr. Carey of Serampore was the lucky recipient of the living treasure, and the poem is supposed to be addressed by him to the dear little flower of his home, thus born under a foreign sky. Dr. Carey was a great lover of flowers, and it was one of his last directions on his death-bed, as I have already said, that his garden should be always protected from the intrusion of Goths and Vandals in the form of Bengallee goats and cows. I must give one stanza of Montgomery's second poetical tribute to the small flower with "the silver crest and golden eye."

> Thrice-welcome, little English flower! To this resplendent hemisphere Where Flora's giant offsprings tower In gorgeous liveries all the year; Thou, only thou, art little here Like worth unfriended and unknown, Yet to my British heart more dear Than all the torrid zone.

It is difficult to exaggerate the feeling with which an exile welcomes a home-flower. A year or two ago Dr. Ward informed the Royal Institution of London, that a single primrose had been taken to Australia in a glass-case and that when it arrived there in full bloom, the sensation it excited was so great that even those who were in the hot pursuit of gold, paused in their eager career to gaze for a moment upon the flower of their native fields, and such immense crowds at last pressed around it that it actually became necessary to protect it by a guard.

My last poetical tribute to the Daisy shall be three stanzas from Wordsworth, from two different addresses to the same flower.

> With little here to do or see Of things that in the great world be, Sweet Daisy! oft I talk to thee, For thou art worthy, Thou unassuming Common-place Of Nature, with that homely face, And yet with something of a grace, Which Love makes for thee!
>
> If stately passions in me burn, And one chance look to Thee should turn, I drink out of an humbler urn A lowlier pleasure; The homely sympathy that heeds The common life, our nature breeds; A wisdom fitted to the needs Of hearts at leisure. When, smitten by the morning ray, I see thee rise,

alert and gay, Then, cheerful Flower! my spirits play With kindred gladness; And when, at dusk, by dews opprest Thou sink'st, the image of thy rest Hath often eased my pensive breast Of careful sadness.

It is peculiarly interesting to observe how the profoundest depths of thought and feeling are sometimes stirred in the heart of genius by the smallest of the works of Nature. Even more ordinarily gifted men are similarly affected to the utmost extent of their intellect and sensibility. We grow tired of the works of man. In the realms of art we ever crave something unseen before. We demand new fashions, and when the old are once laid aside, we wonder that they should ever have excited even a moment's admiration. But Nature, though she is always the same, never satiates us. The simple little Daisy which Burns has so sweetly commemorated is the same flower that was "of all flowres the flowre," in the estimation of the Patriarch of English poets, and which so delighted Wordsworth in his childhood, in his middle life, and in his old age. He gazed on it, at intervals, with unchanging affection for upwards of fourscore years.

The Daisy--the miniature sun with its tiny rays--is especially the favorite of our earliest years. In our remembrances of the happy meadows in which we played in childhood, the daisy's silver lustre is ever connected with the deeper radiance of its gay companion, the butter-cup, which when held against the dimple on the cheek or chin of beauty turns it into a little golden dell. The thoughtful and sensitive frequenter of rural scenes discovers beauty every where; though it is not always the sort of beauty that would satisfy the taste of men who recognize no gaiety or loveliness beyond the walls of cities. To the poet's eye even the freckles on a milk-maid's brow are not without a grace, associated as they are with health, and the open sunshine.

Chaucer tells us that the French call the Daisy *La belle Marguerite*. There is a little anecdote connected with the appellation. Marguerite of Scotland, the Queen of Louis the Eleventh, presented Marguerite Clotilde de Surville, a poetess, with a bouquet of daisies, with this inscription; "Marguerite d'Ecosse à Marguerite (*the pearl*) d'Helicon."

The country maidens in England practise a kind of sortilége with this flower. They pluck off leaf by leaf, saying alternately "*He loves me*" and "*He loves me not.*" The omen or oracle is decided by the fall of either sentence on the last leaf.

It is extremely difficult to rear the daisy in India. It is accustomed to all weathers in England, but the long continued sultriness of this clime makes it as delicate as a languid English lady in a tropical exile, and however carefully and skilfully nursed, it generally pines for its native air and dies.[088]

THE PRICKLY GORSE.

> --Yon swelling downs where the sweet air stirs The harebells, and where prickly furze Buds lavish gold.

Keat's Endymion.

> Fair maidens, I'll sing you a song, I'll tell of the bonny wild flower, Whose blossoms so yellow, and branches so long, O'er moor and o'er rough rocky mountains are flung Far a-way from trim garden and bower

L.A. Tuamley.

The PRICKLY GORSE or Goss or Furze, (*ulex*)[089] I cannot omit to notice, because it was the plant which of all others most struck Dillenius when he first trod on English ground. He threw himself on his knees and thanked Heaven that he had lived to see the golden undulation of acres of wind-waved gorse. Linnaeus lamented that he could scarcely keep it alive in Sweden even in a greenhouse.

I have the most delightful associations connected with this plant, and never think of it without a summer feeling and a crowd of delightful images and remembrances of rural quietude and blue skies and balmy breezes. Cowper hardly does it justice:

> The common, over-grown with fern, and rough With prickly gorse, that shapeless and deformed And dangerous to the touch, has yet its bloom And decks itself with ornaments of gold, Yields no unpleasing ramble.

The plant is indeed irregularly shaped, but it is not *deformed*, and if it is dangerous to the touch, so also is the rose, unless it be of that species which Milton places in Paradise--"*and without thorns the rose.*"

Hurdis is more complimentary and more just to the richest ornament of the swelling hill and the level moor.

> And what more noble than the vernal furze With golden caskets hung?

I have seen whole *cotees* or *coteaux* (sides of hills) in the sweet little island of Jersey thickly mantled with the golden radiance of this beautiful wildflower. The whole Vallée des Vaux (*the valley of vallies*) is sometimes alive with its lustre.

VALLEE DES VAUX.

AIR--THE MEETING OF THE WATERS.

> If I dream of the past, at fair Fancy's command, Up-floats from the blue sea thy small sunny land! O'er thy green hills, sweet Jersey, the fresh breezes blow, And silent and warm is the Vallée des Vaux! There alone have I loitered 'mid blossoms of gold, And forgot that the great world was crowded and cold, Nor believed that a land of enchantment could show A vale more divine than the Vallée des Vaux. A few scattered cots, like white clouds in the sky, Or like still sails at sea when the light breezes die, And a mill with its wheel in the brook's silver glow, Form thy beautiful hamlet, sweet Vallée des Vaux! As the brook prattled by like an infant at play, And each wave as it passed stole a moment away, I thought how serenely a long life would flow, By the sweet little brook in the Vallée des Vaux.

D.L.R.

Jersey is not the only one of the Channel Islands that is enriched with "blossoms of gold." In the sister island of Guernsey the prickly gorse is much used for hedges, and Sir George Head remarks that the premises of a Guernsey farmer are thus as impregnably fortified and secured as if his grounds were surrounded by a stone wall. In the Isle of Man the furze grows so high that it is sometimes more like a fir tree than the ordinary plant.

There is an old proverb:--"When gorse is out of blossom, kissing is out of fashion"--that is *never*. The gorse blooms all the year.

FERN.

> I'll seek the shaggy fern-clad hill And watch, 'mid murmurs muttering stern, The seed departing from the fern Ere wakeful demons can convey The wonder-working charm away.

Leyden.

"The green and graceful Fern" (*filices*) with its exquisite tracery must not be overlooked. It recalls many noble home-scenes to British eyes. Pliny says that "of ferns there are two kinds, and they bear neither flowers nor seed." And this erroneous notion of the fern bearing no seed was common amongst the English even so late as the time of Addison who ridicules "a Doctor that had arrived at the knowledge of the green and red dragon, *and had discovered the female fern-seed*." The seed is very minute and might easily escape a careless eye. In the present day every one knows that the seed of the fern lies on the under side of the leaves, and a single leaf will often bear some millions of seeds. Even those amongst the vulgar who believed the plant bore seed, had an idea that the seeds were visible only at certain mysterious seasons and to favored individuals who by carrying a quantity of it on their person, were able, like those who wore the helmet of Pluto or the ring of Gyges, to walk unseen amidst a crowd. The seed was supposed to be best seen at a certain hour of the night on which St. John the Baptist was born.

> We have the receipt of fern-seed; we walk invisible,

Shakespeare's Henry IV. Part I.

In Beaumont's and Fletcher's *Fair Maid of the Inn,* is the following allusion to the fern.

> --Had you Gyges' ring, *Or the herb that gives invisibility.*

Ben Jonson makes a similar allusion to it:

> I had No medicine, sir, to go invisible, *No fern-seed in my pocket.*

Pope puts a branch of spleen-wort, a species of fern, (*Asplenium trichomanes*) into the hand of a gnome as a protection from evil influences in the Cave of Spleen.

> Safe passed the gnome through this fantastic band A branch of healing spleen-wort in his hand.

The fern forms a splendid ornament for shadowy nooks and grottoes, or fragments of ruins, or heaps of stones, or the odd corners of a large garden or pleasure-ground.

I have had many delightful associations with this plant both at home and abroad. When I visited the beautiful Island of Penang, Sir William Norris, then the Recorder of the Island, and who was a most indefatigable collector of ferns, obligingly presented me with a specimen of every variety that he had discovered in the hills and vallies of that small paradise; and I suppose that in no part of the world could a finer collection of specimens of the fern be made for a botanist's *herbarium*. Fern leaves will look almost as well ten years after they are gathered as on the day on which they are transferred from the dewy hillside to the dry pages of a book.

Jersey and Penang are the two loveliest islands on a small scale that I have yet seen: the latter is the most romantic of the two and has nobler trees and a richer soil and a brighter sky--but they are both charming retreats for the lovers of peace and nature. As I have devoted some verses to Jersey I must have some also on

THE ISLAND OF PENANG.

> I. I stand upon the mountain's brow-- I drink the cool fresh, mountain breeze-- I see thy little town below,[090] Thy villas, hedge-rows, fields and trees, And hail thee with exultant glow, GEM OF THE ORIENTAL SEAS! II. A cloud had settled on my heart-- My frame had borne perpetual pain-- I

yearned and panted to depart From dread Bengala's sultry plain-- Fate smiled,--Disease withholds his dart-- I breathe the breath of life again! III. With lightened heart, elastic tread, Almost with youth's rekindled flame, I roam where loveliest scenes outspread Raise thoughts and visions none could name, Save those on whom the Muses shed A spell, a dower of deathless fame. IV. I *feel*, but oh! could ne'er *pourtray*, Sweet Isle! thy charms of land and wave, The bowers that own no winter day, The brooks where timid wild birds lave, The forest hills where insects gay[091] Mimic the music of the brave! V. I see from this proud airy height A lovely Lilliput below! Ships, roads, groves, gardens, mansions white, And trees in trimly ordered row,[092] Present almost a toy like sight, A miniature scene, a fairy show! VI. But lo! beyond the ocean stream, That like a sheet of silver lies, As glorious as a poet's dream The grand Malayan mountains rise, And while their sides in sunlight beam Their dim heads mingle with the skies. VI. Men laugh at bards who live *in clouds*-- The clouds *beneath* me gather now, Or gliding slow in solemn crowds, Or singly, touched with sunny glow, Like mystic shapes in snowy shrouds, Or lucid veils on Beauty's brow. VIII. While all around the wandering eye Beholds enchantments rich and rare, Of wood, and water, earth, and sky A panoramic vision fair, The dyal breathes his liquid sigh, And magic floats upon the air! IX. Oh! lovely and romantic Isle! How cold the heart thou couldst not please! Thy very dwellings seem to smile Like quiet nests mid summer trees! I leave thy shores--but weep the while-- GEM OF THE ORIENTAL SEAS!

D.L.R.

HENNA.

The henna or al hinna (*Lawsonia inermis*) is found in great abundance in Egypt, India, Persia and Arabia. In Bengal it goes by the name of *Mindee*. It is much used here for garden hedges. Hindu females rub it on the palms of their hands, the tips of their fingers and the soles of their feet to give them a red dye. The same red dye has been observed upon the nails of Egyptian mummies. In Egypt sprigs of henna are hawked about the streets for sale with the cry of

"*O, odours of Paradise; O, flowers of the henna!*" Thomas Moore alludes to one of the uses of the henna:--

> Thus some bring leaves of henna to imbue The fingers' ends of a bright roseate hue, So bright, that in the mirror's depth they seem Like tips of coral branches in the stream.

MOSS.

MOSSES (*musci*) are sometimes confounded with Lichens. True mosses are green, and lichens are gray. All the mosses are of exquisitely delicate structure. They are found in every part of the world where the atmosphere is moist. They have a wonderful tenacity of life and can often be restored to their original freshness after they have been dried for years. It was the sight of a small moss in the interior of Africa that suggested to Mungo Park such consolatory reflections as saved him from despair. He had been stripped of all he had by banditti.

"In this forlorn and almost helpless condition," he says, "when the robbers had left me, I sat for some time looking around me with amazement and terror. Whichever way I turned, nothing appeared but danger and difficulty. I found myself in the midst of a vast wilderness, in the depth of the rainy season--naked and alone,--surrounded by savages. I was five hundred miles from any European settlement. All these circumstances crowded at once upon my recollection; and I confess that my spirits began to fail me. I considered my fate as certain, and that I had no alternative, but to lie down and perish. The influence of religion, however aided and supported me. I reflected that no human prudence or foresight could possibly have averted my present sufferings. I was indeed a stranger in a strange land, yet I was still under the eye of that Providence who has condescended to call himself the stranger's friend. At this moment, painful as my reflections were, the extraordinary beauty of a small Moss irresistibly caught my eye; and though the whole plant was not larger than the top of one of my fingers, I could not contemplate the delicate conformation of its roots, leaves, and fruit, without admiration. Can that Being (thought I) who planted, watered, and brought to perfection, in this obscure part of the world, a thing which appears of so small importance, look with unconcern upon the situation and sufferings of creatures formed after his own image? Surely not.--Reflections like these would not allow me to despair. I started up; and disregarding both, hunger and fatigue, traveled forward, assured that relief was at hand; and I was not disappointed."

VICTORIA REGIA.

On this Queen of Aquatic Plants the language of admiration has been exhausted. It was discovered in the first year of the present century by the botanist Haenke who was sent by the Spanish Government to investigate the vegetable productions of Peru. When in a canoe on the Rio Mamore, one of the great tributaries of the river Amazon, he came suddenly upon the noblest and largest flower that he had ever seen. He fell on his knees in a transport of admiration. It was the plant now known as the Victoria Regia, or American Water-lily.

It was not till February 1849, that Dr. Hugh Rodie and Mr. Lachie of Demerara forwarded seeds of the plant to Sir W.T. Hooker in vials of pure water. They were sown in earth, in pots immersed in water, and enclosed in a glass case. They vegetated rapidly. The plants first came to perfection at Chatsworth the seat of the Duke of Devonshire,[093] and subsequently at the Royal gardens at Kew.

Early in November of the same year, (1849,) the leaves of the plant at Chatsworth were 4 feet 8 inches in diameter. A child weighing forty two pounds was placed upon one of the leaves which bore the weight well. The largest leaf of the plant by the middle of the next month was five feet in diameter with a turned up edge of from two to four inches. It then bore up a person of 11 stone weight. The flat leaf of the Victoria Regia as it floats on the surface of the water, resembles in point of form the brass high edged platter in which Hindus eat their rice.

The flowers in the middle of May 1850 measured one foot one inch in diameter. The rapidity of the growth of this plant is one of its most remarkable characteristics, its leaves often expanding eight inches in diameter daily, and Mr. John Fisk Allen, who has published in America an admirably illustrated work upon the subject, tells us that instances under his own observation have occurred of the leaves increasing at the rate of half an inch hourly.

Not only is there an extraordinary variety in the colours of the several specimens of this flower, but a singularly rapid succession of changes of hue in the same individual flower as it progresses from bud to blossom.

This vegetable wonder was introduced into North America in 1851. It grows to a larger size there than in England. Some of the leaves of the plant cultivated in North America measure seventy-two inches in diameter.

This plant has been proved to be perennial. It grows best in from 4 to 6 feet of water. Each plant generally sends but four or five leaves to the surface.

In addition to the other attractions of this noble Water Lily, is the exquisite character of its perfume, which strongly resembles that of a fresh pineapple just cut open.

The Victoria Regia in the Calcutta Botanic Garden has from some cause or other not flourished so well as it was expected to do. The largest leaf is not more than four feet and three quarters in diameter. But there can be little doubt that when the habits of the plant are better understood it will be brought to great perfection in this country. I strongly recommend my native friends to decorate their tanks with this the most glorious of aquatic plants.

THE FLY-ORCHIS--THE BEE-ORCHIS.

Of these strange freaks of nature many strange stories are told. I cannot repeat them all. I shall content myself with quoting the following passage from D'Israeli's *Curiosities of Literature*:--

"There is preserved in the British Museum, a black stone, on which nature has sketched a resemblance of the portrait of Chaucer. Stones of this kind, possessing a sufficient degree of resemblance, are rare; but art appears not to have been used. Even in plants, we find this sort of resemblance. There is a species of the orchis found in the mountainous parts of Lincolnshire, Kent, &c. Nature has formed a bee, apparently feeding on the breast of the flower, with so much exactness, that it is impossible at a very small distance to distinguish the imposition. Hence the plant derives its name, and is called, the *Bee-flower*. Langhorne elegantly notices its appearance.

> See on that floweret's velvet breast, How close the busy vagrant lies? His thin-wrought plume, his downy breast, Th' ambrosial gold that swells his thighs. Perhaps his fragrant load may bind His limbs;--we'll set the captive free-- I sought the living bee to find, And found the picture of a bee,'

The late Mr. James of Exeter wrote to me on this subject: 'This orchis is common near our sea-coasts; but instead of being exactly like a BEE, *it is not like it at all*. It has a general resemblance to a *fly*, and by the help of imagination, may be supposed to be a fly pitched upon the flower. The mandrake very frequently has a forked root, which may be fancied to resemble thighs and legs. I have seen it helped out with nails on the toes.'

An ingenious botanist, a stranger to me, after reading this article, was so kind as to send me specimens of the *fly* orchis, *ophrys muscifera*, and of the *bee* orchis, *ophrys apifera*. Their resemblance to these insects when in full flower is the most perfect conceivable; they are distinct plants. The poetical eye of Langhorne was equally correct and fanciful; and that too of Jackson, who differed so positively. Many controversies have been carried on, from a want of a little more knowledge; like that of the BEE *orchis* and the FLY *orchis*; both parties prove to be right."[094]

THE FUCHSIA.

The Fuchsia is decidedly the most *graceful* flower in the world. It unfortunately wants fragrance or it would be the *beau ideal* of a favorite of Flora. There is a story about its first introduction into England which is worth reprinting here:

'Old Mr. Lee, a nurseryman and gardener, near London, well known fifty or sixty years ago, was one day showing his variegated treasures to a friend, who suddenly turned to him, and declared, 'Well, you have not in your collection a prettier flower than I saw this morning at Wapping!'--'No! and pray what was this phoenix like?' 'Why, the plant was elegant, and the flowers hung in rows like tassels from the pendant branches; their colour the richest crimson; in the centre a fold of deep purple,' and so forth. Particular directions being demanded and given, Mr. Lee posted off to Wapping, where he at once perceived that the plant was new in this part of the world. He saw and admired. Entering the house, he said, 'My good woman, that is a nice plant. I should like to buy it.'--'I could not sell it for any money, for it was brought me from the West Indies by my husband, who has now left again, and I must keep it for his sake.'--'But I must have it!'--'No sir!'--'Here,' emptying his pockets; 'here are gold, silver, copper.' (His stock was something more than eight guineas.)--'Well a-day! but this is a power of money, sure and sure.'--''Tis yours, and the plant is mine; and, my good dame, you shall have one of the first young ones I rear, to keep for your husband's sake,'--'Alack, alack!'--'You shall.' A coach was called, in which was safely deposited our florist and his seemingly dear purchase. His first work was to pull off and utterly destroy every vestige of blossom and bud. The plant was divided into cuttings, which were forced in bark beds and hotbeds; were redivided and subdivided. Every effort was used to multiply it. By the commencement of the next flowering season, Mr. Lee was the delighted possessor of 300 Fuchsia plants, all giving promise of blossom. The two which opened first were removed into his show-house. A lady came:--'Why, Mr. Lee, my dear Mr. Lee, where did you get this charming flower?'--'Hem! 'tis a new thing, my lady; pretty, is it not?'--'Pretty! 'tis lovely. Its price?'--'A guinea: thank your ladyship;' and one of the plants stood proudly in her ladyship's boudoir. 'My dear Charlotte, where did you get?' &c.--'Oh! 'tis a new thing; I saw it at old Lee's; pretty, is it not?'--'Pretty! 'tis beautiful! Its price!'--'A guinea; there was another left.' The visitor's horses smoked off to the suburb; a third flowering plant stood on the spot whence the first had been taken. The second guinea was paid, and the second chosen Fuchsia adorned the drawing-room of her second ladyship The scene was repeated, as new-comers saw and were attracted by the beauty of the plant. New chariots flew to the gates of old Lee's nursery-ground. Two Fuchsias, young, graceful and bursting into healthy flower, were constantly seen on the same spot in his repository. He neglected not to gladden the faithful sailor's wife by the promised gift; but, ere the flower season closed, 300 golden guineas clinked in his purse, the produce of the single shrub of the widow of Wapping; the reward of the taste, decision, skill, and perseverance of old Mr. Lee.'

Whether this story about the fuchsia, be only partly fact and partly fiction I shall not pretend to determine; but the best authorities acknowledge that Mr. Lee, one of the founders of the Hammersmith Nursery, was the first to make the plant generally known in England and that he for some time got a guinea for each of the cuttings. The fuchsia is a native of Mexico and Chili. I believe that most of the plants of this genus introduced into India have flourished for a brief period and then sickened and died.

The poets of England have not yet sung the Fuschia's praise. Here are three stanzas written for a gentleman who had been presented, by the lady of his love with a superb plant of this kind.

A FUCHSIA.

> I. A deed of grace--a graceful gift--and graceful too the giver! Like ear-rings on thine own fair head, these long buds hang and quiver: Each tremulous taper branch is thrilled--flutter the wing-like leaves-- For thus to part from thee, sweet maid, the floral spirit grieves! II. Rude gods in brass or gold enchant an untaught devotee-- Fair marble shapes, rich paintings old, are Art's idolatry; But nought e'er charmed a human breast like this small tremulous flower, Minute and delicate work divine of world-creative power! III. This flower's the Queen of all earth's flowers, and loveliest things appear Linked by some secret sympathy, in this mysterious sphere; The giver and the gift seem one, and thou thyself art nigh When this glory of the garden greets thy lover's raptured eye.

D.L.R.

"Do you know the proper name of this flower?" writes Jeremy Bentham to a lady-friend, "and the signification of its name? Fuchsia from Fuchs, a German botanist."

ROSEMARY.

> There's rosemary--that's for remembrance: Pray you, love, remember.

Hamlet

> There's rosemarie; the Arabians Justifie (Physitions of exceeding perfect skill) It comforteth the brain and memory.

Chester.

Bacon speaks of heaths of ROSEMARY (*Rosmarinus*[095]) that "will smell a great way in the sea; perhaps twenty miles." This reminds us of Milton's Paradise.

> So lovely seemed That landscape, and of pure, now purer air, Meets his approach, and to the heart inspires Vernal delight and joy, able to drive All sadness but despair. Now gentle gales Fanning their odoriferous wings, dispense Native perfumes, and whisper whence they stole Those balmy spoils. As when to them who sail Beyond the Cape of Hope, and now are past Mozambic, off at sea north east winds blow Sabean odours from the spicy shore Of Araby the blest, with such delay Well pleased they slack their course, and many a league Cheered with the grateful smell, old Ocean smiles.

Rosemary used to be carried at funerals, and worn as wedding favors.

> *Lewis* Pray take a piece of Rosemary *Miramont* I'll wear it, But for the lady's sake, and none of your's!

Beaumont and Fletcher's "Elder Brother."

Rosemary, says Malone, being supposed to strengthen the memory, was the emblem of fidelity in lovers. So in *A Handfull of Pleasant Delites, containing Sundrie New Sonets, 16mo.* 1854:

> Rosemary is for remembrance Between us daie and night, Wishing that I might alwaies have You present in my sight.

The poem in which these lines are found, is entitled, 'A *Nosegay alwaies sweet for Lovers to send for Tokens of Love.*'

Roger Hochet in his sermon entitled *A Marriage Present* (1607) thus speaks of the Rosemary;--"It overtoppeth all the flowers in the garden, boasting man's rule. It helpeth the brain, strengtheneth the memorie, and is very medicinable for the head. Another propertie of the rosemary is, it affects the heart. Let this rosemarinus, this flower of men, ensigne of your wisdom, love, and loyaltie, be carried not only in your hands, but in your hearts and heads."

"Hungary water" is made up chiefly from the oil distilled from this shrub.

I should talk on a little longer about other shrubs, herbs, and flowers, (particularly of flowers) such as the "pink-eyed Pimpernel" (the poor man's weather glass) and the fragrant Violet, ('the modest grace of the vernal year,') the scarlet crested Geranium with its crimpled leaves, and the yellow and purple Amaranth, powdered with gold,

> A flower which once In Paradise, fast by the tree of life Began to bloom,

and the crisp and well-varnished Holly with "its rutilant berries," and the white Lily, (the vestal Lady of the Vale,--"the flower of virgin light") and the luscious Honeysuckle, and the chaste Snowdrop,

> Venturous harbinger of spring And pensive monitor of fleeting years,

and the sweet Heliotrope and the gay and elegant Nasturtium, and a great many other "bonnie gems" upon the breast of our dear mother earth,--but this gossipping book has already extended to so unconscionable a size that I must quicken my progress towards a conclusion[096].

I am indebted to the kindness of Babu Kasiprasad Ghosh, the first Hindu gentlemen who ever published a volume of poems in the English language[097] for the following interesting list of Indian flowers used in Hindu ceremonies. Many copies of the poems of Kasiprasad Ghosh, were sent to the English public critics, several of whom spoke of the author's talents with commendation. The late Miss Emma Roberts wrote a brief biography of him for one of the

London annuals, so that there must be many of my readers at home who will not on this occasion hear of his name for the first time.

A BRIEF ACCOUNT OF INDIAN FLOWERS, COMMONLY USED IN HINDU CEREMONIES.[098]

A'KUNDA (*Calotropis Gigantea*).--A pretty purple coloured, and slightly scented flower, having a sweet and agreeable smell. It is called *Arca* in Sanscrit, and has two varieties, both of which are held to be sacred to Shiva. It forms one of the five darts with which the Indian God of Love is supposed to pierce the hearts of young mortals.[099] Sir William Jones refers to it in his Hymn to Kama Deva. It possesses medicinal properties.[100]

A'PARA'JITA (*Clitoria ternatea*).--A conically shaped flower, the upper part of which is tinged with blue and the lower part is white. Some are wholly white. It is held to be sacred to Durgá.

ASOCA. (*Jonesia Asoca*).--A small yellow flower, which blooms in large clusters in the month of April and gives a most beautiful appearance to the tree. It is eaten by young females as a medicine. It smells like the Saffron.

A'TASHI.--A small yellowish or brown coloured flower without any smell. It is supposed to be sacred to Shiva, and is very often alluded to by the Indian poets. It resembles the flower of the flax or Linum usitatissimum.[101]

BAKA.--A kidney shaped flower, having several varieties, all of which are held to be sacred to Vishnu, and are in consequence used in his worship. It is supposed to possess medicinal virtues and is used by the native doctors.

BAKU'LA (*Mimusops Etengi*).--A very small, yellowish, and fragrant flower. It is used in making garlands and other female ornaments. Krishna is said to have fascinated the milkmaids of Brindabun by playing on his celebrated flute under a *Baku'la* tree on the banks of the Jumna, which is, therefore, invariably alluded to in all the Sanscrit and vernacular poems relating to his amours with those young women.

BA'KASHA (*Justicia Adhatoda*).--A white flower, having a slight smell. It is used in certain native medicines.

BELA (*Jasminum Zambac*).--A fragrant small white flower, in common use among native females, who make garlands of it to wear in their braids of hair. A kind of *uttar* is extracted from this flower, which is much esteemed by natives. It is supposed to form one of the darts of Kama Deva or the God of Love. European Botanists seem to have confounded this flower with the Monika, which they also call the Jasminum Zambac.

BHU'MI CHAMPAKA.--An oblong variegated flower, which shoots out from the ground at the approach of spring. It has a slight smell, and is considered to possess medicinal properties. The great peculiarity of this flower is that it blooms when there is not apparently the slightest trace of the existence of the shrub above ground. When the flower dies away, the leaves make their appearance.

CHAMPA' (*Michelia Champaka*).--A tulip shaped yellow flower possessing a very strong smell.[102] It forms one of the darts of Kama Deva, the Indian Cupid. It is particularly sacred to Krishna.

CHUNDRA MALLIKA' (*Chrysanthemum Indiana*).--A pretty round yellow flower which blooms in winter. The plant is used in making hedges in gardens and presents a beautiful appearance in the cold weather when the blossoms appear.

DHASTU'RA (*Datura Fastuosa*).--A large tulip shaped white flower, sacred to Mahadeva, the third Godhead of the Hindu Trinity. The seeds of this flower have narcotic properties.[103]

DRONA.--A white flower with a very slight smell.

DOPATI (*Impatiens Balsamina*).--A small flower having a slight smell. There are several varieties of this flower. Some are red and some white, while others are both white and red.

GA'NDA' (*Tagetes erecta*).--A handsome yellow flower, which sometimes grows very large. It is commonly used in making garlands, with which the natives decorate their idols, and the Europeans in India their churches and gates on Christmas Day and New Year's Day.

GANDHA RA'J (*Gardenia Florida*).--A strongly scented white flower, which blooms at night.

GOLANCHA (*Menispermum Glabrum*).--A white flower. The plant is already well known to Europeans as a febrifuge.

JAVA' (*Hibiscus Rosa Sinensis*).--A large blood coloured flower held to be especially sacred to Kali. There are two species of it, viz. the ordinary Javá commonly seen in our gardens and parterres, and the *Pancha Mukhi*, which, as its name imports, has five compartments and is the largest of the two.[104]

JAYANTI (*Aeschynomene Sesban*).--A small yellowish flower, held to be sacred to Shiva.

JHA'NTI.--A small white flower possessing medicinal properties. The leaves of the plants are used in curing certain ulcers.

JA'NTI (*Jasminum Grandiflorum*).--Also a small white flower having a sweet smell. The *uttar* called *Chumeli* is extracted from it.

JUYIN (*Jasminum Auriculatum*).--The Indian Jasmine. It is a very small white flower remarkable for its sweetness. It is also used in making a species of *uttar* which is highly prized by the natives, as also in forming a great variety of imitation female ornaments.

KADAMBA (*Nauclea Cadamba*).--A ball shaped yellow flower held to be particularly sacred to Krishna, many of whose gambols with the milkmaids of Brindabun are said to have been performed under the Kadamba tree, which is in consequence very frequently alluded to in the vernacular poems relating to his loves with those celebrated beauties.

KINSUKA (*Butea Frondosa*).--A handsome but scentless white flower.

KANAKA CHAMPA (*Pterospermum Acerifolium*).--A yellowish flower which hangs down in form of a tassel. It has a strong smell, which is perceived at a great distance when it is on the tree, but the moment it is plucked off, it begins to lose its fragrance.

KANCHANA (*Bauhinia Variegata*).--There are several varieties of this flower. Some are white, some are purple, while others are red. It gives a handsome appearance to the tree when the latter is in full blossom.

KUNDA (*Jasminum pulescens*).--A very pretty white flower. Indian poets frequently compare a set of handsome teeth, to this flower. It is held to be especially sacred to Vishnu.

KARABIRA (*Nerium Odosum*).--There are two species of this flower, viz. the white and red, both of which are sacred to Shiva.

KAMINI (*Murraya Exotica*).--A pretty small white flower having a strong smell. It blooms at night and is very delicate to the touch. The *kamini* tree is frequently used as a garden hedge.

KRISHNA CHURA (*Poinciana Pulcherrima*).--A pretty small flower, which, as its name imports resembles the head ornament of Krishna. When the Krishna Chura tree is in full blossom, it has a very handsome appearance.

KRISHNA KELI (*Mirabilis Jalapa.*)[105]--A small tulip shaped yellow flower. The bulb of the plant has medicinal properties and is used by the natives as a poultice.

KUMADA (*Nymphaea Esculenta*)--A white flower, resembling the lotus, but blooming at night, whence the Indian poets suppose that it is in love with Chandra or the Moon, as the lotus is imagined by them to be in love with the Sun.

LAVANGA LATA' (*Limonia Scandens.*)--A very small red flower growing upon a creeper, which has been celebrated by Jaya Deva in his famous work called the *Gita Govinda*. This creeper is used in native gardens for bowers.

MALLIKA' (*Jasminum Zambac.*)--A white flower resembling the *Bela*. It has a very sweet smell and is used by native females to make ornaments. It is frequently alluded to by Indian poets.

MUCHAKUNDA (*Pterospermum Suberifolia*).--A strongly scented flower, which grows in clusters and is of a brown colour.

MA'LATI (*Echites Caryophyllata.*)--The flower of a creeper which is commonly used in native gardens. It has a slight smell and is of a white colour.

MA'DHAVI (*Gaertnera Racemosa.*)--The flower of another creeper which is also to be seen in native gardens. It is likewise of a white colour.

NA'GESWARA (*Mesua Ferrua*.)--A white flower with yellow filaments, which are said to possess medicinal properties and are used by the native physicians. It has a very sweet smell and is supposed by Indian poets to form one of the darts of Kama Deva. See Sir William Jones's Hymn to that deity.

PADMA (*Nelumbium Speciosum*.)--The Indian lotus, which is held to be sacred to Vishnu, Brama, Mahadava, Durga, Lakshami and Saraswati as well as all the higher orders of Indian deities. It is a very elegant flower and is highly esteemed by the natives, in consequence of which the Indian poets frequently allude to it in their writings.

PA'RIJATA (*Buchanania Latifolia*.)--A handsome white flower, with a slight smell. In native poetry, it furnishes a simile for pretty eyes, and is held to be sacred to Vishnu.

PAREGATA (*Erythrina Fulgens*.)--A flower which is supposed to bloom in the garden of Indra in heaven, and forms the subject of an interesting episode in the *Puranas*, in which the two wives of Krisna, (Rukmini and Satyabhama) are said to have quarrelled for the exclusive possession of this flower, which their husband had stolen from the celestial garden referred to. It is supposed to be identical with the flower of the *Palta madar*.

RAJANI GANDHA (*Polianthus Tuberosa*.)--A white tulip-shaped flower which blooms at night, from which circumstance it is called "the Rajani Gandha, (or night-fragrance giver)." It is the Indian tuberose.

RANGANA.--A small and very pretty red flower which is used by native females in ornamenting their betels.

SEONTI. *Rosa Glandulefera*. A white flower resembling the rose in size and appearance. It has a sweet smell.

SEPHA'LIKA (*Nyctanthes Arbor-tristis*.)--A very pretty and delicate flower which blooms at night, and drops down shortly after. It has a sweet smell and is held to be sacred to Shiva. The juice of the leaves of the Sephalika tree are used in curing both remittant and intermittent fevers.

SURYJA MUKHI (*Helianthus Annuus*).--A large and very handsome yellow flower, which is said to turn itself to the Sun, as he goes from East to West, whence it has derived its name.

SURYJA MANI (*Hibiscus Phoeniceus*).--A small red flower.

GOLAKA CHAMPA.--A large beautiful white tulip-shaped flower having a sweet smell. It is externally white but internally orange-colored.

TAGUR (*Tabernoemontana Coronaria*).--A white flower having a slight smell.

TARU LATA.--A beautiful creeper with small red flowers. It is used in native gardens for making hedges.

K.G.

Pliny in his Natural History alludes to the marks of time exhibited in the regular opening and closing of flowers. Linnaeus enumerates forty- six flowers that might be used for the construction of a floral time- piece. This great Swedish botanist invented a Floral horologe, "whose wheels were the sun and earth and whose index-figures were flowers." Perhaps his invention, however, was not wholly original. Andrew Marvell in his "*Thoughts in a Garden*" mentions a sort of floral dial:--

> How well the skilful gardener drew Of flowers and herbs this dial new! Where, from above, the milder sun Does through a fragrant zodiac run: And, as it works, th'industrious bee Computes its time as well as we: How could such sweet and wholesome hours Be reckoned, but with herbs and flowers?

Marvell[106]

Milton's notation of time--"*at shut of evening flowers*," has a beautiful simplicity, and though Shakespeare does not seem to have marked his time on a floral clock, yet, like all true poets, he has made very free use of other appearances of nature to indicate the commencement and the close of day.

> The sun no sooner shall the mountains touch-- Than we will ship him hence.

Hamlet.

> Fare thee well at once! The glow-worm shows the matin to be near And gins to pale his uneffectual fire.

Hamlet.

> But look! The morn, in russet mantle clad, Walks o'er the dew of yon high eastern hill:-- Break we our watch up.

Hamlet.

> *Light thickens*, and the crow Makes wing to the rooky wood.

Macbeth.

Such picturesque notations of time as these, are in the works of Shakespeare, as thick as autumnal leaves that strew the brooks in Valombrosa. In one of his Sonnets he thus counts the years of human life by the succession of the seasons.

> To me, fair friend, you never can be old, For as you were when first your eye I eyed, Such seems your beauty still. Three winters cold Have from the forests shook three summers' pride; Three beauteous springs to yellow autumn turned In process of the seasons have I seen; Three April's perfumes in three hot Junes burned Since first I saw you fresh which yet are green.

Grainger, a prosaic verse-writer who once commenced a paragraph of a poem with "Now, Muse, let's sing of rats!" called upon the slave drivers in the West Indies to time their imposition of cruel tasks by the opening and closing of flowers.

> Till morning dawn and Lucifer withdraw His beamy chariot, let not the loud bell Call forth thy negroes from their rushy

couch: And ere the sun with mid-day fervor glow, When every broom-bush opes her yellow flower, Let thy black laborers from their toil desist: Nor till the broom her every petal lock, Let the loud bell recal them to the hoe, But when the jalap her bright tint displays, When the solanum fills her cup with dew, And crickets, snakes and lizards gin their coil, Let them find shelter in their cane-thatched huts.

Sugar Cane.[107]

I shall here give (*from Loudon's Encyclopaedia of Gardening*) the form of a flower dial. It may be interesting to many of my readers:--

'Twas a lovely thought to mark the hours As they floated in light away By the opening and the folding flowers That laugh to the summer day.[108]

Mr. Hemans.

A FLOWER DIAL.

TIME OF OPENING.

	[109]	h.	m.
YELLOW GOAT'S BEARD	T.P.	3	5
LATE FLOWERING DANDELION	Leon.S.	4	0
BRISTLY HELMINTHIA	H.B.	4	5
ALPINE BORKHAUSIA	B.A.	4	5
WILD SUCCORY	C.I.	4	5
NAKED STALKED POPPY	P.N.	5	0
COPPER COLOURED DAY LILY	H.F.	5	0
SMOOTH SOW THISTLE	S.L.	5	0

		h.	m.
ALPINE AGATHYRSUS	Ag.A.	5	0
SMALL BIND WEED	Con.A.	5	6
COMMON NIPPLE WORT	L.C.	5	6
COMMON DANDELION	L.T.	5	6
SPORTED ACHYROPHORUS	A.M.	6	7
WHITE WATER LILY	N.A.	7	0
GARDEN LETTUCE	Lec.S.	7	0
AFRICAN MARIGOLD	T.E.	7	0
COMMON PIMPERNEL	A.A.	7	8
MOUSE-EAR HAWKWEED	H.P.	8	0
PROLIFEROUS PINK	D.P.	8	0
FIELD MARIGOLD	Cal.A.	9	0
PURPLE SANDWORT	A.P.	9	10
SMALL PURSLANE	P.O.	9	10
CREEPING MALLOW	M.C.	9	10
CHICKWEED	S.M.	9	10

TIME OF CLOSING.

		h.	m.
HELMINTHIA ECHIOIDES	B.H.	12	0
AGATHYRSUS ALPINUS	A.B.	12	0
BORKHAUSIA ALPINA	A.B.	12	0

LEONTODON SEROTINUS	L.D.	12	0
MALVA CAROLINIANA	C.M.	12	1
DAINTHUS PROLIFER	P.P.	1	0
HIERACIUM PILOSELLA	M.H.	0	2
ANAGALLIS ARVENSIS	S.P.	2	3
ARENARIA PURPUREA	P.S.	2	4
CALENDULA ARVENSIS	F.M.	3	0
TACETES ERECTA	A.M.	3	3
CONVOLVULUS ARVENSIS	S.B.	4	0
ACHYROPHORUS MACULATUS	S.A.	4	5
NYMPHAEA ALBA	W.W.B.	5	0
PAPAVER NUDICAULE	N.P.	7	0
HEMEROCALLIS FULVA	C.D.L.	7	0
CICHORIUM INTYBUS	W.S.	8	9
TRAGOPOGON PRATENSIS	Y.G.B.	9	10
STELLARIA MEDIA	C.	9	10
LAPSANA COMMUNIS	C.N.	10	0
LACTUCA SATIVA	G.L.	10	0
SONCHUS LAEVIS	S.T.	11	10
PORTULACA OLERACEA	S.P.	11	12

Of course it will be necessary to adjust the *Horologium Florae* (or Flower clock) to the nature of the climate. Flowers expand at a later hour in a cold climate than in a warm one. "A flower," says Loudon, "that opens at six o'clock in the morning at Senegal, will not open in France or England till eight or nine, nor in Sweden till ten. A flower that opens at ten o'clock at Senegal will not open in France or England till noon or later, and in Sweden it will not open at all. And a flower that does not open till noon or later at Senegal will not open at all in France or England. This seems as if heat or its absence were also (as well as light) an agent in the opening and shutting of flowers; though the opening of such as blow only in the night cannot be attributed to either light or heat."

The seasons may be marked in a similar manner by their floral representatives. Mary Howitt quotes as a motto to her poem on *Holy Flowers* the following example of religious devotion timed by flowers:--

"Mindful of the pious festivals which our church prescribes," (says a Franciscan Friar) "I have sought to make these charming objects of floral nature, the *time-pieces of my religious calendar*, and the mementos of the hastening period of my mortality. Thus I can light the taper to our Virgin Mother on the blowing of the white snowdrop which opens its floweret at the time of Candlemas; the lady's smock and the daffodil, remind me of the Annunciation; the blue harebell, of the Festival of St George; the ranunculus, of the Invention of the Cross; the scarlet lychnis, of St. John the Baptist's day; the white lily, of the Visitation of our Lady, and the Virgin's bower, of her Assumption; and Michaelmas, Martinmas, Holyrood, and Christmas, have all their appropriate monitors. I learn the time of day from the shutting of the blossoms of the Star of Jerusalem and the Dandelion, and the hour of the night by the stars."

Some flowers afford a certain means of determining the state of the atmosphere. If I understand Mr. Tyas rightly he attributes the following remarks to Hartley Coleridge.--

"Many species of flowers are admirable barometers. Most of the bulbous- rooted flowers contract, or close their petals entirely on the approach of rain. The African marigold indicates rain, if the corolla is closed after seven or eight in the morning. The common bindweed closes its flowers on the approach of rain; but the anagallis arvensis, or scarlet pimpernel, is the most sure in its indications as the petals constantly close on the least humidity of the atmosphere.

Barley is also singularly affected by the moisture or dryness of the air. The awns are furnished with stiff points, all turning towards one end, which extend when moist, and shorten when dry. The points, too, prevent their receding, so that they are drawn up or forward; as moisture is returned, they advance and so on; indeed they may be actually seen to travel forwards. The capsules of the geranium furnish admirable barometers. Fasten the beard, when fully ripe, upon a stand, and it will twist itself, or untwist, according as the air is moist or dry. The flowers of the chick-weed, convolvulus, and oxalis, or wood sorrel, close their petals on the approach of rain."

The famous German writer, Jean Paul Richter, describes what he calls *a Human Clock*.

A HUMAN CLOCK.

"I believe" says Richter "the flower clock of Linnaeus, in Upsal (*Horologium Florae*) whose wheels are the sun and earth, and whose index-figures are flowers, of which one always awakens and opens later than another, was what secretly suggested my conception of the human clock.

I formerly occupied two chambers in Scheeraw, in the middle of the market place: from the front room I overlooked the whole market-place and the royal buildings and from the back one, the botanical garden. Whoever now dwells in these two rooms possesses an excellent harmony, arranged to his hand, between the flower clock in the garden and the human clock in the marketplace. At three o'clock in the morning, the yellow meadow goats-beard opens; and brides awake, and the stable-boy begins to rattle and feed the horses beneath the lodger. At four o'clock the little hawk weed awakes, choristers going to the Cathedral who are clocks with chimes, and the bakers. At five, kitchen maids, dairy maids, and butter-cups awake. At six, the sow-thistle and cooks. At seven o'clock many of the Ladies' maids are awake in the Palace, the Chicory in my botanical garden, and some tradesmen. At eight o'clock all the colleges awake and the little mouse-ear. At nine o'clock, the female nobility already begin to stir; the marigold, and even many young ladies, who have come from the country on a visit, begin to look out of their windows. Between ten and eleven o'clock the Court Ladies and the whole staff of Lords of the Bed-chamber, the green colewort and the Alpine dandelion, and the reader of the Princess rouse themselves out of their morning sleep; and the whole Palace, considering that the morning sun gleams so brightly to-day from the lofty sky through the coloured silk curtains, curtails a little of its slumber.

At twelve o'clock, the Prince: at one, his wife and the carnation have their eyes open in their flower vase. What awakes late in the afternoon at four o'clock is only the red-hawkweed, and the night watchman as cuckoo-clock, and these two only tell the time as evening-clocks and moon-clocks.

From the eyes of the unfortunate man, who like the jalap plant (Mirabilia jalapa), first opens them at five o'clock, we will turn our own in pity aside. It is a rich man who only exchanges the fever fancies of being pinched with hot pincers for waking pains.

I could never know when it was two o'clock, because at that time, together with a thousand other stout gentlemen and the yellow mouse-ear, I always fell asleep; but at three o'clock in the afternoon, and at three in the morning, I awoke as regularly as though I was a repeater. Thus we mortals may be a flower-clock for higher beings, when our flower-leaves close upon our last bed; or sand clocks, when the sand of our life is so run down that it is renewed in the other world; or picture-clocks because, when our death-bell here below strikes and rings, our image steps forth, from its case into the next world.

On each event of the kind, when seventy years of human life have passed away, they may perhaps say, what! another hour already gone! how the time flies!"--*From Balfour's Phyto-Theology*.

Some of the natives of India who possess extensive estates might think it worth their while to plant a LABYRINTH for the amusement of their friends. I therefore give a plan of one from London's *Arboretum et Fruticetum Britannicum*. It would not be advisable to occupy much of a limited estate in a toy of this nature; but where the ground required for it can be easily spared or would otherwise be wasted, there could be no objection to adding this sort of amusement to the very many others that may be included in a pleasure ground. The plan here given, resembles the labyrinth at Hampton

Court. The hedges should be a little above a man's height and the paths should be just wide enough for two persons abreast. The ground should be kept scrupulously clean and well rolled and the hedges well trimmed, or in this country the labyrinth would soon be damp and unwholesome, especially in the rains. To prevent its affording a place of refuge and concealment for snakes and other reptiles, the gardener should cut off all young shoots and leaves within half a foot of the ground. The centre building should be a tasteful summer-house, in which people might read or smoke or take refreshments. To make the labyrinth still more intricate Mr. Loudon suggests that stop-hedges might be introduced across the path, at different places, as indicated in the figure by dotted lines.[110]

A GARDEN LABYRINTH

Of strictly Oriental trees and shrubs and flowers, perhaps the majority of Anglo Indians think with much less enthusiasm than of the common weeds of England. The remembrance of the simplest wild flower of their native fields will make them look with perfect indifference on the decorations of an Indian Garden. This is in no degree surprising. Yet nature is lovely in all lands.

Indian scenery has not been so much the subject of description in either prose or verse as it deserves, but some two or three of our Anglo-Indian authors have touched upon it. Here is a pleasant and truthful passage from an article entitled "*A Morning Walk in India*," written by the late Mr. Lawson, the Missionary, a truly good and a highly gifted man:--

"The rounded clumps that afford the deepest shade, are formed by the mangoe, the banian, and the cotton trees. At the verge of this deep- green forest are to be seen the long and slender hosts of the betle and cocoanut trees; and the grey bark of their trunks, as they catch the light of the morning, is in clear relief from the richness of the back- ground. These as they wave their feathery tops, add much to the picturesque interest of the straw-built hovels beneath them, which are variegated with every tinge to be found amongst the browns and yellows, according to the respective periods of their construction. Some of them are enveloped in blue smoke, which oozes through every interstice of the thatch, and spreads itself, like a cloud hovering over these frail habitations, or moves slowly along, like a strata of vapour not far from the ground, as though too heavy to ascend, and loses itself in the thin air, so inspiring to all who have courage to leave their beds and enjoy it. The champa tree forms a beautiful object in this jungle. It may be recognized immediately from the surrounding scenery. It has always been a favourite with me. I suppose most persons, at times, have been unaccuntably attracted by an object comparatively trifling in itself. There are also particular seasons, when the mind is susceptible of peculiar impressions, and the moments of happy, careless youth, rush upon the imagination with a thousand tender feelings. There are few that do not recollect with what pleasure they have grasped a bunch of wild flowers, when, in the days of their childhood, the languor of a lingering fever has prevented them for some weary months from enjoying that chief of all the pleasures of a robust English boy, a ramble through the fields, where every tree, and bush, and hillock, and blossom, are endeared to him, because, next to a mother's caresses, they were the first things in the world upon which he opened his eyes, and, doubtless, the first which gave him those indescribable feelings of fairy pleasure, which even in his dreams were excited; while the coloured clouds of heaven, the golden sunshine of a landscape, the fresh nosegay of dog-roses and early daisies, and the sounds of busy whispering trees and tinkling brooks presented to the sleeping child all the pure pleasure of his waking moments. And who is there here that does not sometimes recal some of those feelings which were his solace perhaps thirty years ago? Should I be wrong, were I to say that even, at his desk, amid all the excitements and anxieties of commercial pursuits, the weary Calcutta merchant has been lulled into a sort of pensive reminiscence of the past, and, with his pen placed between his lips and his fevered forehead leaning upon his hand, has felt his heart bound at some vivid picture rising upon his imagination. The forms of a fond mother, and an almost angel-looking sister, have been so strongly conjured up with the scenes of his boyish days, that the pen has been unceremoniously dashed to the ground, and 'I will go home' was the sigh that heaved from a bosom full of kindness and English feeling; while, as the dream vanished, plain truth told its tale, and the man of commerce is still to be seen at his desk, pale, and getting into years and perhaps less desirous than ever of winding up his concern. No wonder! because the dearest ties of his heart have been broken, and those who were the charm of home have gone down to the cold grave, the home of all. Why then should he revisit his native place? What is the cottage of his birth to him? What charms has the village now for the gentleman just arrived from India? Every well remembered object of nature, seen after a lapse of twenty years, would only serve to renew a host of buried, painful feelings. Every visit to the house of a surviving neighbour would but bring to mind some melancholy incident; for into what house could he enter, to idle away an hour, without seeing some wreck of his own family, such as a venerable clock, once so loved for the painted moon that waxed and waned to the astonishment of the gazer, or some favorite ancient chair, edged so nobly with rows of brass nails,

--but perforated sore, and dull'd in holes By worms voracious, eating through and through.

These are little things, but they are objects which will live in his memory to the latest day of his life, and with which are associated in his mind the dearest feelings and thoughts of his happiest hours."

Here is an attempt at a description in verse of some of the most common

TREES AND FLOWERS OF BENGAL

> This land is not my father land, And yet I love it--for the hand Of God hath left its mark sublime On nature's face in every clime-- Though from home and friends we part, Nature and the human heart Still may soothe the wanderer's care-- And his God is every where Beneath BENGALA'S azure skies, No vallies sink, no green hills rise, Like those the vast sea billows make-- The land is level as a lake[111] But, oh, what giants of the wood Wave their wide arms, or calmly brood Each o'er his own deep rounded shade When noon's fierce sun the breeze hath laid, And all is still. On every plain How green the sward, or rich the grain! In jungle wild and garden trim, And open lawn and covert dim, What glorious shrubs and flowerets gay, Bright buds, and lordly beasts of prey! How prodigally Gunga pours Her wealth of waves through verdant shores O'er which the sacred peepul bends, And oft its skeleton lines extends Of twisted root, well laved and bare, Half in water, half in air! Fair scenes! where breeze and sun diffuse The sweetest odours, fairest hues-- Where brightest the bright day god shows, And where his gentle sister throws Her softest spell on silent plain, And stirless wood, and slumbering main-- Where the lucid starry sky Opens most to mortal eye The wide and mystic dome serene Meant for visitants unseen, A dream like temple, air built hall, Where spirits pure hold festival! Fair scenes! whence envious Art might steal More charms than fancy's realms reveal-- Where the tall palm to the sky Lifts its wreath triumphantly-- And the bambu's tapering bough Loves its flexile arch to throw-- Where sleeps the favored lotus white, On the still lake's bo-

som bright-- Where the champac's[112] blossoms shine, Offerings meet for Brahma's shrine, While the fragrance floateth wide O'er velvet lawn and glassy tide-- Where the mangoe tope bestows Night at noon day--cool repose, Neath burning heavens--a hush profound Breathing o'er the shaded ground-- Where the medicinal neem, Of palest foliage, softest gleam, And the small leafed tamarind Tremble at each whispering wind-- And the long plumed cocoas stand Like the princes of the land, Near the betel's pillar slim, With capital richly wrought and trim-- And the neglected wild sonail Drops her yellow ringlets pale-- And light airs summer odours throw From the bala's breast of snow-- Where the Briarean banyan shades The crowded ghat, while Indian maids, Untouched by noon tide's scorching rays, Lave the sleek limb, or fill the vase With liquid life, or on the head Replace it, and with graceful tread And form erect, and movement slow, Back to their simple dwellings go-- [Walls of earth, that stoutly stand, Neatly smoothed with wetted hand-- Straw roofs, yellow once and gay, Turned by time and tempest gray--] Where the merry minahs crowd Unbrageous haunts, and chirrup loud-- And shrilly talk the parrots green 'Midst the thick leaves dimly seen-- And through the quivering foliage play, Light as buds, the squirrels gay, Quickly as the noontide beams Dance upon the rippled streams-- Where the pariah[113] howls with fear, If the white man passeth near-- Where the beast that mocks our race With taper finger, solemn face, In the cool shade sits at ease Calm and grave as Socrates-- Where the sluggish buffaloe Wallows in mud--and huge and slow, Like massive cloud of sombre van, Moves the land leviathan--[114] Where beneath the jungle's screen Close enwoven, lurks unseen The couchant tiger--and the snake His sly and sinuous way doth make Through the rich mead's grassy net, Like a miniature rivulet-- Where small white cattle, scattered wide, Browse, from dawn to even tide-- Where the river watered soil Scarce demands the ryot's toil-- And the rice field's emerald light Out vies Italian meadows bright,-- Where leaves of every shape and dye, And blossoms varied as the sky, The fancy kindle,--fingers fair That never closed on aught but air-- Hearts, that never heaved a sigh--

Wings, that never learned to fly-- Cups, that ne'er went table round-- Bells, that never rang with sound-- Golden crowns, of little worth-- Silver stars, that strew the earth-- Filagree fine and curious braid, Breathed, not labored, grown, not made-- Tresses like the beams of morn Without a thought of triumph worn-- Tongues that prate not--many an eye Untaught midst hidden things to pry-- Brazen trumpets, long and bright, That never summoned to the fight-- Shafts, that never pierced a side-- And plumes that never waved with pride;-- Scarcely Art a shape may know But Nature here that shape can show. Through this soft air, o'er this warm sod, Stern deadly Winter never trod; The woods their pride for centuries wear, And not a living branch is bare; Each field for ever boasts its bowers, And every season brings its flowers.

D.L.R.

We all "uphold Adam's profession": we are all gardeners, either practically or theoretically. The love of trees and flowers, and shrubs and the green sward, with a summer sky above them, is an almost universal sentiment. It may be smothered for a time by some one or other of the innumerable chances and occupations of busy life; but a painting in oils by Claude or Gainsborough, or a picture in words by Spenser or Shakespeare that shall for ever

> Live in description and look green in song,

or the sight of a few flowers on a window-sill in the city, can fill the eye with tears of tenderness, or make the secret passion for nature burst out again in sudden gusts of tumultuous pleasure and lighten up the soul with images of rural beauty. There are few, indeed, who, when they have the good fortune to escape on a summer holiday from the crowded and smoky city and find themselves in the heart of a delicious garden, have not a secret consciousness within them that the scene affords them a glimpse of a true paradise below. Rich foliage and gay flowers and rural quiet and seclusion and a smiling sun are ever associated with ideas of earthly felicity.

> And oh, if there be an Elysium on earth, It is this, it is this!

The princely merchant and the petty trader, the soldier and the sailor, the politician and the lawyer, the artist and the artisan, when they pause for a moment in the midst of their career, and dream of the happiness of some future day, almost invariably fix their imaginary palace or cottage of delight in a garden, amidst embowering trees and fragrant flowers. This disposition, even in the busiest men, to indulge occasionally in fond anticipations of rural bliss--

In visions so profuse of pleasantness--

shows that God meant us to appreciate and enjoy the beauty of his works. The taste for a garden is the one common feeling that unites us all.

One touch of nature makes the whole world kin.

There is this much of poetical sensibility--of a sense of natural beauty--at the core of almost every human heart. The monarch shares it with the peasant, and Nature takes care that as the thirst for her society is the universal passion, the power of gratifying it shall be more or less within the reach of all.[115]

Our present Chief Justice, Sir Lawrence Peel, who has set so excellent an example to his countrymen here in respect to Horticultural pursuits and the tasteful embellishment of what we call our *"compounds"* and who, like Sir William Jones and Sir Thomas Noon Talfourd, sees no reason why Themis should be hostile to the Muses, has obliged me with the following stanzas on the moral or rather religious influence of a garden. They form a highly appropriate and acceptable contribution to this volume.

I HEARD THY VOICE IN THE GARDEN.

> That voice yet speaketh, heed it well-- But not in tones of wrath it chideth, The moss rose, and the lily smell Of God--in them his voice abideth. There is a blessing on the spot The poor man decks--the sun delighteth To smile upon each homely plot, And why? The voice of God inviteth. God knows that he is worshipped there, The chaliced cowslip's graceful bending Is mute devotion, and the air Is sweet with

incense of her lending. The primrose, aye the children's pet, Pale bride, yet proud of its uprooting, The crocus, snowdrop, violet And sweet-briar with its soft leaves shooting. There nestles each--a Preacher each-- (Oh heart of man! be slow to harden) Each cottage flower in sooth doth teach God walketh with us in the garden.

I am surprized that in this city (of Calcutta) where so many kinds of experiments in education have been proposed, the directors of public instruction have never thought of attaching tasteful Gardens to the Government Colleges--especially where Botany is in the regular course of Collegiate studies. The Company's Botanic Garden being on the other side of the river and at an inconvenient distance from the city cannot be much resorted to by any one whose time is precious. An attempt was made not long ago to have the Garden of the Horticultural Society (now forming part of the Company's Botanic Garden) on this side of the river, but the public subscriptions that were called for to meet the necessary expenses were so inadequate to the purpose that the money realized was returned to the subscribers, and the idea relinquished, to the great regret of many of the inhabitants of Calcutta who would have been delighted to possess such a place of recreation and instruction within a few minutes' drive.

Hindu students, unlike English boys in general, remind us of Beattie's Minstrel:--

> The exploit of strength, dexterity and speed To him nor vanity, nor joy could bring.

A sort of Garden Academy, therefore, full of pleasant shades, would be peculiarly suited to the tastes and habits of our Indian Collegians. They are not fond of cricket or leap-frog. They would rejoice to devote a leisure hour to pensive letterings in a pleasure-garden, and on an occasional holiday would gladly pursue even their severest studies, book in hand, amidst verdant bowers. A stranger from Europe beholding them, in their half-Grecian garments, thus wandering amidst the trees, would be reminded of the disciples of Plato.

"It is not easy," observes Lord Kames, "to suppress a degree of enthusiasm, when we reflect on the advantages of gardening with respect to virtuous education. In the beginning of life the deepest impressions are made; and it is a sad truth, that the young student, familiarized to the dirtiness and disorder of many colleges pent within narrow bounds in populous cities, is rendered in a measure insensible to the elegant beauties of art and nature. It seems to me far from an exaggeration, that good professors are not more essential to a college, than a spacious garden, sweetly ornamented, but without any thing glaring or fantastic, is upon the whole to inspire our youth with a taste no less for simplicity than for elegance. In this respect the University of Oxford may justly be deemed a model."

It may be expected that I should offer a few hints on the laying out of gardens. Much has been said (by writers on ornamental and landscape gardening) on *art* and *nature*, and almost always has it been implied that these must necessarily be in direct opposition. I am far from being of this opinion. If art and nature be not in some points of view almost identical, they are at least very good friends, or may easily be made so. They are not necessarily hostile. They admit of the most harmonious combinations. In no place are such combinations more easy or more proper than in a garden. Walter Scott very truly calls a garden the child of Art. But is it not also the child of Nature?--of Nature and Art together? To attempt to exclude art--or even, the appearance of art-- from a small garden enclosure, is idle and absurd. He who objects to all art in the arrangement of a flower-bed, ought, if consistent with himself, to turn away with an expression of disgust from a well arranged nosegay in a rich porcelain vase. But who would not loathe or laugh at such manifest affectation or such thoroughly bad taste? As there is a time for every thing, so also is there a place for every thing. No man of true judgment would desire to trace the hand of human art on the form of nature in remote and gigantic forests, and amidst vast mountains, as irregular as the billows of a troubled sea. In such scenery there is a sublime grace in wildness,--*there* "the very weeds are beautiful." But what true judgment would be enchanted with weeds and wildness in the small parterre. As Pope rightly says, we must

Consult the genius of the place in all.

It is pleasant to enter a rural lane overgrown with field-flowers, or to behold an extensive common irregularly decorated with prickly gorse or fern and thistle, but surely no man of taste would admire nature in this wild and dishevelled state in a little suburban garden. Symmetry, elegance and beauty, (--no *sublimity* or *grandeur*--) trimness, snugness, privacy, cleanliness, comfort, and convenience--the results of a happy conjunction of art and nature--are all that we can aim at within a limited extent of ground. In a small parterre we either trace with pleasure the marks of the gardener's attention or are disgusted with his negligence. In a mere patch of earth around a domestic dwelling nature ought not to be left entirely to herself.

What is agreeable in one sphere of life is offensive in another. A dirty smock frock and a soiled face in a ploughman's child who has been swinging on rustic gates a long summer morning or rolling down the slopes of hills, or grubbing in the soil of his small garden, may remind us, not unpleasantly, of one of Gainsborough's pictures; but we look for a different sort of nature on the canvas of Sir Joshua Reynolds or Sir Thomas Lawrence, or in the brilliant drawing-rooms of the nobility; and yet an Earl's child looks and moves at least as *naturally* as a peasant's.

There is nature every where--in the palace as well as in the hut, in the cultivated garden as well as in the wild wood. Civilized life is, after all, as natural as savage life. All our faculties are natural, and civilized man cultivates his mental powers and studies the arts of life by as true an instinct as that which leads the savage to make the most of his mud hut, and to improve himself or his child as a hunter, a fisherman, or a warrior. The mind of man is the noblest work of its Maker (--in this world--) and the movements of man's mind may be quite as natural, and quite as poetical too, as the life that rises from the ground. It is as natural for the mind, as it is for a tree or flower to advance towards perfection. Nature suggests art, and art again imitates and approximates to nature, and this principle of action and reaction brings man by degrees towards that point of comparative excellence for which God seems to have intended him. The mind of a Milton or a Shakespeare is surely not in a more unnatural condition than that of an ignorant rustic. We ought not then to decry refinement nor deem all connection of art with nature an offensive incongruity. A noble mansion in a spacious and well kept park is an object which even an observer who has no share himself in the property may look upon with pleasure. It makes him proud of his race.[116] We cannot witness so harmonious a conjunction of art and nature without feeling that man is something better than a mere beast of the field or forest. We see him turn both art and nature to his service, and we cannot contemplate the lordly dwelling and the richly decorated land around it--and the neatness and security and order of the whole scene--without associating them with the high accomplishments and refined tastes that in all probability distinguish the proprietor and his family. It is a strange mistake to suppose that nothing is natural beyond savage ignorance--that all refinement is unnatural--that there is only one sort of simplicity. For the mind elevated by civilization is in a more natural state than a mind that has scarcely passed the boundary of brutal instinct, and the simplicity of a savage's hut, does not prevent there being a nobler simplicity in a Grecian temple.

Kent[117] the famous landscape gardener, tells us that *natureabhors a straight line*. And so she does--in some cases--but not in all. A ray of light is a straight line, and so also is a Grecian nose, and so also is the stem of the betel-nut tree. It must, indeed, be admitted that he who should now lay out a large park or pleasure-ground on strictly geometrical principles or in the old topiary style would exhibit a deplorable want of taste and judgment. But the provinces of the landscape gardener and the parterre gardener are perfectly distinct. The landscape gardener demands a wide canvas. All his operations are on a large scale. In a small garden we have chiefly to aim at the *gardenesque* and in an extensive park at the *picturesque*. Even in the latter case, however, though

'Tis Nature still, 'tis nature methodized:

Or in other words:

Nature to advantage dressed.

for even in the largest parks or pleasure-grounds, an observer of true taste is offended by an air of negligence or the absence of all traces of human art or care. Such places ought to indicate the presence of civilized life and security and order. We are not pleased to see weeds and jungle--or litter of any sort--even dry leaves--upon the princely domain, which should look like a portion of nature set apart or devoted to the especial care and enjoyment of the owner and his friends:--a strictly private property. The grass carpet should be trimly shorn and well swept. The trees should be tastefully separated from each other at irregular but judicious distances. They should have fine round heads of foliage, clean stems, and no weeds or underwood below, nor a single dead branch above. When we visit the finest estates of the nobility and gentry in England it is impossible not to perceive in every case a marked distinction between the wild nature of a wood and the civilized nature of a park. In the latter you cannot overlook the fact that every thing injurious to the health and growth and beauty of each individual tree has been studiously removed, while on the other hand, light, air, space, all things in fact that, if sentient, the tree could itself be supposed to desire, are most liberally supplied. There is as great a difference between the general aspect of the trees in a nobleman's pleasure ground and those in a jungle, as between the rustics of a village and the well bred gentry of a great city. Park trees have generally a fine air of aristocracy about them.

A Gainsborough or a Morland would seek his subjects in remote villages and a Watteau or a Stothard in the well kept pleasure ground. The ruder nature of woods and villages, of sturdy ploughmen and the healthy though soiled and ragged children in rural neighbourhoods, affords a by no means unpleasing contrast and introduction to the trim trees and smoothly undulating lawns, and curved walks, and gay parterres, and fine ladies and well dressed and graceful children on some old ancestral estate. We look for rusticity in the village, and for elegance in the park. The sleek and noble air of patrician trees, standing proudly on the rich velvet sward, the order and grace and beauty of all that meets the eye, lead us, as I have said already, to form a high opinion of the owner. In this we may of course be sometimes disappointed; but a man's character is generally to be traced in almost every object around him over which he has the power of a proprietor, and in few things are a man's taste and habits more distinctly marked than in his park and garden. If we find the owner of a neatly kept garden and an elegant mansion slovenly, rude and vulgar in appearance and manners, we inevitably experience that shock of surprize which is excited by every thing that is incongruous or out of keeping. On the other hand if the garden be neglected and overgrown with weeds, or if every thing in its arrangement indicate a want of taste, and a disregard of neatness and order, we feel no astonishment whatever in discovering that the proprietor is as negligent of his mind and person as of his shrubberies and his lawns.

A civilized country ought not to look like a savage one. We need not have wild nature in front of our neatly finished porticos. Nothing can be more strictly artificial than all architecture. It would be absurd to erect an elegantly finished residence in the heart of a jungle. There should be an harmonious gradation from the house to the grounds, and true taste ought not to object to terraces of elegant design and graceful urns and fine statues in the immediate neighbourhood of a noble dwelling.

Undoubtedly as a general rule, the undulating curve in garden scenery is preferable to straight lines or abrupt turns or sharp angles, but if there should happen to be only a few yards between the outer gateway and the house, could anything be more fantastical or preposterous than an attempt to give the ground between them a serpentine irregularity? Even in the most spacious grounds the walks should not seem too studiously winding, as if the short turns were meant for no other purpose than to perplex or delay the walker.[118] They should have a natural sweep, and seem to meander rather in accordance with the nature of the ground and the points to which they lead than in obedience to some idle sport of fancy. They should not remind us of Gray's description of the divisions of an old mansion:

> Long passages that lead to nothing.

Foot-paths in small gardens need not be broader than will allow two persons to walk abreast with ease. A spacious garden may have walks of greater breadth. A path for one person only is inconvenient and has a mean look.

I have made most of the foregoing observations in something of a spirit of opposition to those Landscape gardeners who I think once carried a true principle to an absurd excess. I dislike, as much as any one can, the old topiary style of our remote ancestors, but the talk about free nature degenerated at last into downright cant, and sheer extravagance; the reformers were for bringing weeds and jungle right under our parlour windows, and applied to an acre of ground those rules of Landscape gardening which required a whole county for their proper exemplification. It is true that Milton's Paradise had "no nice art" in it, but then it was not a little suburban pleasure ground but a world. When Milton alluded to private gardens, he spoke of their trimness.

> Retired Leisure That in *trim* gardens takes his pleasure.

The larger an estate the less necessary is it to make it merely neat, and symmetrical, especially in those parts of the ground that are distant from the house; but near the architecture some degree of finish and precision is always necessary, or at least advisable, to prevent the too sudden contrast between the straight lines and artificial construction of the dwelling and the flowing curves and wild but beautiful irregularities of nature unmoulded by art. A garden adjacent to the house should give the owner a sense of *home*. He should not feel himself abroad at his own door. If it were only for the sake of variety there should be some distinction between the private garden and the open field. If the garden gradually blends itself with a spacious park or chase, the more the ground recedes from the house the more it may legitimately assume the aspect of a natural landscape. It will then be necessary to appeal to the eye of a landscape gardener or a painter or a poet before the owner, if ignorant of the principles of fine art, attempt the completion of the general design.

I should like to see my Native friends who have extensive grounds, vary the shape of their tanks, but if they dislike a more natural form of water, irregular or winding, and are determined to have them with four sharp corners, let them at all events avoid the evil of several small tanks in the same "compound." A large tank is more likely to have good water and to retain it through the whole summer season than a smaller one and is more easily kept clean and grassy to the water's edge. I do not say that it would be proper to have a piece of winding water in a small compound--that indeed would be impracticable. But even an oval or round tank would be better than a square one.[119]

If the Native gentry could obtain the aid of tasteful gardeners, I would recommend that the level land should be varied with an occasional artificial elevation, nicely sloped or graduated; but Native *malees* would be sure to aim rather at the production of abrupt round knobs resembling warts or excrescences than easy and natural undulations of the surface.

With respect to lawns, the late Mr. Speede recommended the use of the *doob* grass, but it is so extremely difficult to keep it clear of any intermixture of the *ooloo* grass, which, when it intrudes upon the *doob* gives the lawn a patchwork and shabby look, that it is better to use the *ooloo* grass only, for it is far more manageable; and if kept well rolled and closely shorn it has a very neat, and indeed, beautiful appearance. The lawns in the compound of the Government House in Calcutta are formed of *ooloo* glass only, but as they have been very carefully attended to they have really a most brilliant and agreeable aspect. In fact, their beautiful bright green, in the hottest summer, attracts even the notice and admiration of the stranger fresh from England. The *ooloo* grass, however, on close inspection is found to be extremely coarse, nor has even the finest *doob* the close texture and velvet softness of the grass of English lawns.

Flower beds should be well rounded. They should never have long narrow necks or sharp angles in which no plant can have room to grow freely. Nor should they be divided into compartments, too minute or numerous, for so arranged they must always look petty and toy-like. A lawn should be as open and spacious as the ground will fairly admit without too greatly limiting the space for flowers. Nor should there be an unnecessary multiplicity of walks. We should aim at a certain breadth of style. Flower beds may be here and there distributed over the lawn, but care should be taken that it be not too much broken up by them. A few trees may be introduced upon the lawn, but they must not be placed so close together as to prevent the growth of the grass by obstructing either light or air. No large trees should be allowed to smother up the house, particularly on the southern and western sides, for besides impeding the circulation through the rooms of the most wholesome winds of this country, they would attract mosquitoes, and give an air of gloominess to the whole place.

Natives are too fond of over-crowding their gardens with trees and shrubs and flowers of all sorts, with no regard to individual or general effects, with no eye to arrangement of size, form or color; and in this hot and moist climate the consequent exclusion of free air and the necessary degree of light has a most injurious influence not only upon the health of the resident but upon vegetation itself. Neither the finest blossoms nor the finest fruits can be expected from an overstocked garden. The native malee generally plants his fruit trees so close together that they impede each other's growth and strength. Every Englishman when he enters a native's garden feels how much he could improve its productiveness and beauty by a free use of the hatchet. Too many trees and too much embellishment of a small garden make it look still smaller, and even on a large piece of ground they produce confused and disagreeable effects and indicate an absence of all true judgment. This practice of over-filling a garden is an instance of bad taste, analogous to that which is so conspicuously characteristic of our own countrymen in India with respect to their apartments, which look more like an upholsterer's show-rooms or splendid ornament-shops than drawing-rooms or parlours. There is scarcely space enough to turn in them without fracturing some frail and costly bauble. Where a garden is over-planted the whole place is darkened, the ground is green and slimy, the grass thin, sickly and straggling, and the trees and shrubs deficient in freshness and vigor.

Not only should the native gentry avoid having their flower-borders too thickly filled,--they should take care also that they are not too broad. We ought not to be obliged to leave the regular path and go across the soft earth of the bed to obtain a sight of a particular shrub or flower. Close and entangled foliage keeps the ground too damp, obstructs wholesome air, and harbours snakes and a great variety of other noxious reptiles. Similar objections suggest the propriety of having no shrubs or flowers or even a grass-plot immediately under the windows and about the doors of the house. A well exposed gravel or brick walk should be laid down on all sides of the house, as a necessary safeguard against both moisture and vermin.

I have spoken already of the unrivalled beauty of English gravel. It cannot be too much admired. *Kunkur*[120] looks extremely smart for a few weeks while it preserves its solidity and freshness, but it is rapidly ground into powder under carriage wheels or blackened by occasional rain and the permanent moisture of low grounds when only partially exposed to the sun and air. Why should not an opulent Rajah or Nawaub send for a cargo of beautiful red gravel from the gravel pits at Kensington? Any English House of Agency here would obtain it for him. It would be cheap in the end, for it lasts at least five times as long as the kunkur, and if of a proper depth admits of repeated turnings with the spade, looking on every turn almost as fresh as the day on which it was first laid down.

Instead of brick-bat edgings, the wealthy Oriental nobleman might trim all his flower-borders with the green box-plant of England, which would flourish I suppose in this climate or in any other. Cobbett in his *English Gardener* speaks with so much enthusiasm and so much to the purpose on the subject of box as an edging, that I must here repeat his eulogium on it.

The box is at once the most efficient of all possible things, and the prettiest plant that can possibly be conceived; the color of its leaf; the form of its leaf; its docility as to height, width and shape; the compactness of its little branches; its great durability as a plant; its thriving in all sorts of soils and in all sorts of aspects; *its freshness under the hottest sun*, and its defiance of all shade and drip: these are the beauties and qualities which, for ages upon ages, have marked it out as the chosen plant for this very important purpose.

The edging ought to be clipped in the winter or very early in spring on both sides and at top; a line ought to be used to regulate the movements of the shears; it ought to be clipped again in the same manner about midsummer; and if there be *a more neat and beautiful thing than this in the world, all that I can say is, that I never saw that thing.*

A small green edging for a flower bed can hardly be too *trim*; but large hedges with tops and sides cut as flat as boards, and trees fantastically shaped with the shears into an exhibition as full of incongruities as the wildest dream, have deservedly gone out of fashion in England. Poets and prose writers have agreed to ridicule all verdant sculpture on a large scale. Here is a description of the old topiary gardens.

> These likewise mote be seen on every side The shapely box, of all its branching pride Ungently shorn, and, with preposterous skill To various beasts, and birds of sundry quill Transformed, and human shapes of monstrous size.
>
> Also other wonders of the sportive shears Fair Nature misadorning; there were found Globes, spiral columns, pyramids, and piers With spouting urns and budding statues crowned; And horizontal dials on the ground In living box, by cunning artists traced, And galleys trim, or on long voyage bound, But by their roots there ever anchored fast.

G. West.

The same taste for torturing nature into artificial forms prevailed amongst the ancients long after architecture and statuary had been carried to such perfection that the finest British artists of these times can do nothing but copy and repeat what was accomplished so many ages ago by the people of another nation. Pliny, in his description of his Tuscan villa, speaks of some of his trees having been cut into letters and the forms of animals, and of others placed in such regular order that they reminded the spectator of files of soldiers.[121] The Dutch therefore should not bear all the odium of the topiary style of gardening which they are said to have introduced into England and other countries of Europe. They were not the first sinners against natural taste.

The Hindus are very fond of formally cut hedges and trimmed trees. All sorts of verdant hedges are in some degree objectionable in a hot moist country, rife with deadly vermin. I would recommend ornamental iron railings or neatly cut and well painted wooden pales, as more airy, light, and cheerful, and less favorable to snakes and centipedes.

This is the finest country in the world for making gardens speedily. In the rainy season vegetation springs up at once, as at the stroke of an Enchanter's wand. The Landscape gardeners in England used to grieve that they could hardly expect to live long enough to see the effect of their designs. Such artists would have less reason, to grieve on that account in this country. Indeed even in England, the source of uneasiness alluded to, is now removed. "The deliberation with which trees grow," wrote Horace Walpole, in a letter to a friend, "is extremely inconvenient to my natural impatience. I lament living in so barbarous an age when we are come to so little perfection in gardening. I am persuaded that 150 years hence it will be as common to remove oaks 150 years old as it now is to plant tulip roots." The writer was not a bad prophet. He has not yet been dead much more than half a century and his expectations are already more than half realized. Shakespeare could not have anticipated this triumph of art when he made Macbeth ask

> Who can impress the forest? Bid the tree Unfix his earthbound root?

The gardeners have at last discovered that the largest (though not perhaps the *oldest*) trees can be removed from one place to another with comparative facility and safety. Sir H. Stewart moved several hundred lofty trees without the least injury to any of them. And if broad and lofty trees can be transplanted in England, how much more easily and securely might such a process be effected in the rainy season in this country. In half a year a new garden might be made to look like a garden of half a century. Or an old and ill-arranged plantation might thus be speedily re-adjusted to the taste of the owner. The main object is to secure a good ball of earth round the root, and the main difficulty is to raise the tree and remove it. Many most ingenious machines for raising a tree from the ground, and trucks for removing it, have been lately invented by scientific gardeners in England. A Scotchman, Mr. McGlashen, has been amongst the most successful of late transplanters. He exhibited one of his machines at Paris to the present Emperor of the French, and lifted with it a fir tree thirty feet high. The French ruler lavished the warmest commendations on the ingenious artist and purchased his apparatus at a large price.[122]

Bengal is enriched with a boundless variety of noble trees admirably suited to parks and pleasure grounds. These should be scattered about a spacious compound with a spirited and graceful irregularity, and so disposed with reference to the dwelling as in some degree to vary the view of it, and occasionally to conceal it from the visitor driving up the winding road from the outer gate to the portico. The trees, I must repeat, should be so divided as to give them a free growth and admit sufficient light and air beneath them to allow the grass to flourish. Grassless ground under park trees has a look of barrenness, discomfort and neglect, and is out of keeping with the general character of the scene.

The Banyan (*Ficus Indica or Bengaliensis*)--

> The Indian tree, whose branches downward bent, Take root again, a boundless canopy--

and the Peepul or Pippul (*Ficus Religiosa*) are amongst the finest trees in this country--or perhaps in the world--and on a very spacious pleasure ground or park they would present truly magnificent aspects. Colonel Sykes alludes to a Banyan at the village of Nikow in Poonah with 68 stems descending from and supporting the branches. This tree is said to be capable of affording shelter to 20,000 men. It is a tree of this sort which Milton so well describes.

> The fig tree, not that kind for fruit renowned, But such as at this day, to Indians known In Malabar or Deccan, spreads her arms Branching so broad and long, a pillared shade, High over arched, and echoing walks between There oft the Indian herdsman, shunning heat, Shelters in cool, and tends his pasturing herds At loop holes cut through the thickest shade those leaves, They gathered, broad as Amazonian taige; And with what skill they had together sewed, To gird their waste.

Milton is mistaken as to the size of the leaves of this tree, though he has given its general character with great exactness.[123]

A remarkable banyan or buri tree, near Manjee, twenty miles west of Patna, is 375 inches in diameter, the circumference of its shadow at noon measuring 1116 feet. It has sixty stems, or dropped branches that have taken root. Under this tree once sat a naked fakir who had occupied that situation for 25 years; but he did not continue there the whole year, for his vow obliged him to be during the four cold months up to his neck in the water of the Ganges![124]

It is said that there is a banyan tree near Gombroon on the Persian gulf, computed to cover nearly 1,700 yards.

The Banyan tree in the Company's Botanic garden, is a fine tree, but it is of small dimensions compared with those of the trees just mentioned.[125]

The cocoanut tree has a characteristically Oriental aspect and a natural grace, but it is not well suited to the ornamental garden or the princely villa. It is too suggestive of the rudest village scenery, and perhaps also of utilitarian ideas of mere profit, as every poor man who has half a dozen cocoanut trees on his ground disposes of the produce in the bazar.

I would recommend my native friends to confine their clumps of plaintain trees to the kitchen garden, for though the leaf of the plaintain is a proud specimen of oriental foliage when it is first opened out to the sun, it soon gets torn to shreds by the lightest breeze. The tattered leaves then dry up and the whole of the tree presents the most beggarly aspect imaginable. The stem is as ragged and untidy as the leaves.

The kitchen garden and the orchard should be in the rear of the house. The former should not be too visible from the windows and the latter is on many accounts better at the extremity of the grounds than close to the house, as we too often find it. A native of high rank should keep as much out of sight as possible every thing that would remind a visitor that any portion of the ground was intended rather for pecuniary profit than the immediate pleasure of the owner. The people of India do not seem to be sufficiently aware that any sign of parsimony in the management of a large park or pleasure ground produces in the mind of the visitor an unfavorable impression of the character of the owner. I have seen in Calcutta vast mansions of which every little niche and corner towards the street was let out to very small traders at a few annas a month. What would the people of England think of an opulent English Nobleman who should try to squeeze a few pence from the poor by dividing the street front of his palace into little pigeon-sheds of petty shops for the retail of petty wares? Oh! Princes of India "reform this altogether." This sordid saving, this widely published parsimony, is not only not princely, it is not only not decorous, it is positively disgusting to every passer-by who himself possesses any right thought or feeling.

The Natives seem every day more and more inclined to imitate European fashions, and there are few European fashions, which could be borrowed by the highest or lowest of the people of this country with a more humanizing and delightful effect than that attention to the exterior elegance and neatness of the dwelling-house, and that tasteful garniture of the contiguous ground, which in England is a taste common to the prince and the peasant, and which has made that noble country so full of those beautiful homes which surprize and enchant its foreign visitors.

The climate and soil of this country are peculiarly favorable to the cultivation of trees and shrubs and flowers; and the garden here is at no season of the year without its ornaments.

The example of the Horticultural Society of India, and the attractions of the Company's Botanic Garden ought to have created a more general taste amongst us for the culture of flowers. Bishop Heber tells us that the Botanic Garden here reminded hint more of Milton's description of the Garden of Eden than any other public garden, that he had ever seen.[126]

There is a Botanic Garden at Serampore. In 1813 it was in charge of Dr. Roxburgh. Subsequently came the amiable and able Dr. Wallich; then the venerable Dr. Carey was for a time the Officiating Superintendent. Dr. Voigt followed and then one of the greatest of our Anglo-Indian botanists, Dr. Griffiths. After him came Dr. McLelland, who is at this present time counting the teak trees in the forests of Pegu. He was succeeded by Dr. Falconer who left this country but a few months ago. The garden is now in charge of Dr.

Thomson who is said to be an enthusiast in his profession. He explored the region beyond the snowy range I think with Captain Cunningham, some years ago. With the exceptions of Voigt and Carey, all who have had charge of the garden at Serampore have held at the same time the more important appointment of Superintendent of the Company's Botanic Garden at Garden Beach.

There is a Botanic Garden at Bhagulpore, which owes its origin to Major Napleton. I have been unable to obtain any information regarding its present condition. A good Botanic Garden has been already established in the Punjab, where there is also an Agricultural and Horticultural Society.

I regret that it should have been deemed necessary to make stupid pedants of Hindu malees by providing them with a classical nomenclature for plants. Hindostanee names would have answered the purpose just as well. The natives make a sad mess of our simplest English names, but their Greek must be Greek indeed! A *Quarterly Reviewer* observes that Miss Mitford has found it difficult to make the maurandias and alstraemerias and eschxholtzias--the commonest flowers of our modern garden--look passable even in prose. But what are these, he asks, to the pollopostemonopetalae and eleutheroromacrostemones of Wachendorf, with such daily additions as the native name of iztactepotzacuxochitl icohueyo, or the more classical ponderosity of Erisymum Peroffskyanum.

> --like the verbum Graecum Spermagoraiolekitholakanopolides, Words that should only be said upon holidays, When one has nothing else to do.

If these names are unpronounceable even by Europeans, what would the poor Hindu malee make of them? The pedantry of some of our scientific Botanists is something marvellous. One would think that a love of flowers must produce or imply a taste for simplicity and nature in all things.[127]

As by way of encouragement to the native gardeners--to enable them to dispose of the floral produce of their gardens at a fair price--the Horticultural Society has withdrawn from the public the indulgence of gratuitous supplies of plants, it would be as well if some men of taste were to instruct these native nursery-men how to lay out their grounds, (as their fellow-traders do at home,) with some regard to neatness, cleanliness and order. These flower-merchants, and even the common *malees*, should also be instructed, I think, how to make up a decent bouquet, for if it be possible to render the most elegant things in the creation offensive to the eye of taste, that object is assuredly very completely effected by these swarthy artists when they arrange, with such worse than Dutch precision and formality, the ill-selected, ill- arranged, and tightly bound treasures of the parterre for the classical vases of their British masters. I am often vexed to observe the idleness or apathy which suffers such atrocities as these specimens of Indian taste to disgrace the drawing-rooms of the City of Palaces. This is quite inexcusable in a family where there are feminine hands for the truly graceful and congenial task of selecting and arranging the daily supply of garden decorations. A young lady--"herself a fairer flower"-- is rarely exhibited to a loving eye in a more delightful point of view than when her delicate and dainty fingers are so employed.

If a lovely woman arranging the nosegays and flower-vases, in her parlour, is a sweet living picture, a still sweeter sight does she present to us when she is in the garden itself. Milton thus represents the fair mother of the fair in the first garden:--

> Eve separate he spies. Veil'd in a cloud of fragrance, where she stood, Half spied, so thick the roses blushing round About her glow'd, oft stooping to support Each flower of slender stalk, whose head, though gay, Carnation, purple, azure, or speck'd with gold, Hung drooping unsustain'd; them she upstays Gently with myrtle band, mindless the while Herself, though fairest unsupported flower, From her best prop so far, and storm so nigh. Nearer he drew, and many a walk traversed Of stateliest covert, cedar, pine, or palm; Then voluble and bold, now hid, now seen, Among thick woven arborets, and flowers Imborder'd on each bank, the hand of Eve[128]

Paradise Lost. Book IX.

Chaucer (in "The Knight's Tale,") describes Emily in her garden as fairer to be seen

> Than is the lily on his stalkie green;

And Dryden, in his modernized version of the old poet, says,

> At every turn she made a little stand, And thrust among the thorns her lily hand To draw the rose.

Eve's roses were without thorns--

> "And without thorn the rose,"[129]

It is pleasant to see flowers plucked by the fairest fingers for some elegant or worthy purpose, but it is not pleasant to see them *wasted*. Some people pluck them wantonly, and then fling them away and litter the garden walks with them. Some idle coxcombs, vain

> Of the nice conduct of a clouded cane,

amuse themselves with switching off their lovely heads. "That's villainous, and shows a most pitiful ambition in the fool that uses it." Lander says

> And 'tis my wish, and over was my way, To let all flowers live freely, and so die.

Here is a poetical petitioner against a needless destruction of the little tenants of the parterre.

> Oh, spare my flower, my gentle flower, The slender creature of a day, Let it bloom out its little hour, And pass away. So soon its fleeting charms must lie Decayed, unnoticed and o'erthrown, Oh, hasten not its destiny, Too like thine own.

Lyte.

Those who pluck flowers needlessly and thoughtlessly should be told that other people like to see them flourish, and that it is as well for every one to bear in mind the beautiful remark of Lord Bacon that "the breath of flowers is far sweeter in the air than in the hand; for in the air it comes and goes like the warbling of music."

The British portion of this community allow their exile to be much more dull and dreary than it need be, by neglecting to cultivate their gardens, and leaving them entirely to the taste and industry of the *malee*. I never feel half so much inclined to envy the great men of this now crowded city the possession of vast but gardenless mansions, (partly blocked up by those of their neighbours,) as I do to felicitate the owner of some humbler but more airy and wholesome dwelling in the suburbs, when the well-sized grounds attached to it have been touched into beauty by the tasteful hand of a lover of flowers.

But generally speaking my countrymen in most parts of India allow their grounds to remain in a state which I cannot help characterizing as disreputable. It is amazing how men or women accustomed to English modes of life can reconcile themselves to that air of neglect, disorder, and discomfort which most of their "compounds" here exhibit.

It would afford me peculiar gratification to find this book read with interest by my Hindu friends, (for whom, chiefly, it has been written,) and to hear that it has induced some of them to pay more attention to the ornamental cultivation of their grounds; for it would be difficult to confer upon them a greater blessing than a taste for the innocent and elegant pleasures of the FLOWER-GARDEN.

SUPPLEMENT.

SACRED TREES AND SHRUBS OF THE HINDUS.

The following list of the trees and shrubs held sacred by the Hindus is from the friend who furnished me with the list of Flowers used in Hindu ceremonies.[130] It was received too late to enable me to include it in the body of the volume.

AMALAKI (*Phyllanthus emblica*).--A tree held sacred to Shiva. It has no flowers, and its leaves are in consequence used in worshipping that deity as well as Durga, Kali, and others. The natives of Bengal do not look upon it with any degree of religious veneration, but those of the Upper Provinces annually worship it on the day of the *Shiva Ratri*, which generally falls in the latter end of February or the beginning of March, and on which all the public offices are closed.

ASWATH-THA (*Ficus Religiosa*).--It is commonly called by Europeans the Peepul tree, by which name, it is known to the natives of the Upper Provinces. The *Bhagavat Gita* says that Krishna in giving an account of his power and glory to Arjuna, before the commencement of the celebrated battle between the *Kauravas* and *Pándavas* at *Kurukshetra*, identified himself with the *Aswath-tha* whence the natives consider it to be a sacred tree.[131]

BILWA OR SREEFUL (*Aegle marmelos*).--It is the common wood-apple tree, which is held sacred to Shiva, and its leaves are used in worshipping him as well as Durga, Kali, and others. The *Mahabharat* says that when Shiva at the request of Krishna and the Pandavas undertook the protection of their camp at Kurukshetra on the night of the last day of the battle, between them and the sons of Dhritarashtra, Aswathama, a friend and follower of the latter, took up a Bilwa tree by its roots and threw it upon the god, who considering it in the light of an offering made to him, was so much pleased with Aswathama that he allowed him to enter the camp, where he killed the five sons of the Pandavas and the whole of the remnants of their army. Other similar stories are also told of the Bilwa tree to prove its sacredness, but the one I have given above, will be sufficient to shew in what estimation it is held by the Hindus.

BAT (*Ficus indica*).--Is the Indian Banian tree, supposed to be immortal and coeval with the gods; whence it is venerated as one of them. It is also supposed to be a male tree, while the Aswath-tha or Peepul is looked upon as a female, whence the lower orders of the people plant them side by side and perform the ceremony of matrimony with a view to connect them as man and wife.[132]

DURVA' (*Panicum dactylon*).--A grass held to be sacred to Vishnu, who in his seventh *Avatara* or incarnation, as Rama, the son of Dasaratha, king of Oude, assumed the colour of the grass, which is used in all religious ceremonies of the Hindus. It has medicinal properties.

KA'STA' (*Saccharum spontaneum*).--It is a large species of grass. In those ceremonies which the Hindus perform after the death of a person, or with a view to propitiate the Manes of their ancestors this grass is used whenever the Kusa is not to be had. When it is in flower, the natives look upon the circumstance as indicative of the close of the rains.

KU'SA (*Poa cynosuroides*).--The grass to which, reference has been made above. It is used in all ceremonies performed in connection with the death of a person or having for their object the propitiation of the Manes of ancestors.

MANSA-SHIJ (*Euphorbia ligularia*).--This plant is supposed by the natives of Bengal to be sacred to *Mansa*, the goddess of snakes, and is worshipped by them on certain days of the months of June, July, August, and September, during which those reptiles lay their eggs and breed their young. The festival of Arandhana, which is more especially observed by the lower orders of the people, is in honor of the Goddess Mansa.[133]

NA'RIKELA (*Coccos nucifera*).--The Cocoanut tree, which is supposed to possess the attributes of a Brahmin and is therefore held sacred.[134]

NIMBA (*Melia azadirachta*).--A tree from the trunk of which the idol at Pooree was manufactured, and which is in consequence identified with the ribs of Vishnu.[135]

TU'LSI (*Ocymum*).--The Indian Basil, of which there are several species, such as the *Ram Tulsi* (ocymum gratissimum) the *Babooye Tulsi* (ocymum pilosum) the *Krishna Tulsi* (osymum sanctum) and the common *Tulsi* (ocymum villosum) all of which possess medicinal properties, but the two latter are held to be sacred to Vishnu and used in his worship. The *Puranas* say that Krishna assumed the form of *Saukasura*, and seduced his wife Brinda. When he was discovered he manifested his extreme regard for her by turning her into the *Tulsi* and put the leaves upon his head.[136]

APPENDIX.

THE FLOWER GARDEN IN INDIA.

The following practical directions and useful information respecting the Indian Flower-Garden, are extracted from the late Mr. Speede's *New Indian Gardener*, with the kind permission of the publishers, Messrs. Thacker Spink and Company of Calcutta.

THE SOIL.

So far as practicable, the soil should be renewed every year, by turning in vegetable mould, river sand, and well rotted manure to the depth of about a foot; and every second or third year the perennials should be taken up, and reduced, when a greater proportion of manure may be added, or what is yet better, the whole of the old earth removed, and new mould substituted.

It used to be supposed that the only time for sowing annuals or other plants, (in Bengal) is the beginning of the cold weather, but although this is the case with a great number of this class of plants, it is a popular error to think it applies to all, since there are many that grow more luxuriantly if sown at other periods. The Pink, for instance, may be sown at any time, Sweet William thrives best if sown in March or April, the variegated and light colored Larkspurs should not be put in until December, the Dahlia germinates most successfully in the rains, and the beautiful class of Zinnias are never seen to perfection unless sown in June.

This is the more deserving of attention, as it holds out the prospect of maintaining our Indian flower gardens, in life and beauty, throughout the whole year, instead of during the confined period hitherto attempted.

The several classes of flowering plants are divided into PERENNIAL, BIENNIAL, and ANNUAL.

PERENNIALS.

The HERON'S BILL, Erodium; the STORK'S BILL, Pelargonium; and the CRANE'S BILL, Geranium; all popularly known under the common designation of Geranium, which gives name to the family, are well known, and are favorite plants, of which but few of the numerous varieties are found in this country.

Of the first of these there are about five and twenty fixed species, besides a vast number of varieties; of which there are here found only the following:--

The *Flesh-colored Heron's bill*, E. incarnatum, is a pretty plant of about six inches high, flowering in the hot weather, with flesh-colored blossoms, but apt to become rather straggling.

Of the hundred and ninety species of the second class, independently of their varieties, there are few indeed that have found their way here, only thirteen, most of which are but rarely met with.

The *Rose-colored Stork's bill*, P. roseum, is tuberous rooted, and in April yields pretty pink flowers.

The *Brick-colored Stork's bill*, P. lateritium, affords red flowers in March and April.

The *Botany Bay Stork's bill*, P. Australe, is rare, but may be made to give a pretty red flower in March.

The *Common horse-shoe Stork's bill*, P. zonale, is often seen, and yields its scarlet blossoms freely in April.

The *Scarlet-flowered Stork's bill*, P. inquinans, affords a very fine flower towards the latter end of the cold weather, and approaching to the hot; it requires protection from the rains, as it is naturally of a succulent nature, and will rot at the joints if the roots become at all sodden: many people lay the pots down on their sides to prevent this, which is tolerably successful to their preservation.

The *Sweet-Scented Stork's bill*, P. odoratissimum, with pink flowers, but it does not blossom freely, and the branches are apt to grow long and straggling.

The *Cut-leaved Stork's bill*, P. incisum, has small flowers, the petals being long and thin, and the flowers which appear in April are white, marked with pink.

The *Ivy-leaved Stork's bill*, P. lateripes, has not been known to yield flowers in this country.

The *Rose-scented Stork's bill*, P. capitatum, the odour of the leaves is very pleasant, but it is very difficult to force into blossom.

The *Ternate Stork's bill*, P. ternatum, has variegated pink flowers in April.

The *Oak-leaved Stork's bill*, P. quercifolium, is much esteemed for the beauty of its leaves, but has not been known to blossom in this climate.

The *Tooth-leaved Stork's bill*, P. denticulatum, is not a free flowerer, but may with care be made to bloom in April.

The *Lemon, or Citron-scented Stork's bill*, P. gratum, grows freely, and has a pretty appearance, but does not blossom.

Of the second class of these plants the forty-eight species have only three representatives.

The *Aconite-leaved Crane's bill*, G. aconiti-folium, is a pretty plant, but rare, yielding its pale blue flowers with difficulty.

The *Wallich's Crane's bill* G. Wallichianum, indigenous to Nepal, having pale pink blossoms and rather pretty foliage, flowering in March and April; but requiring protection in the succeeding hot weather, and the beginning of the rains, as it is very susceptible of heat, or excess of moisture.

Propagation--may be effected by seed to multiply, or produce fresh varieties, but the ordinary mode of increasing the different sorts is by cuttings, no plant growing more readily by this mode. These should be taken off at a joint where the wood is ripening, at which point the root fibres are formed, and put into a pot with a compost of one part garden mould, one part vegetable mould, and one part sand, and then kept moderately moist, in the shade, until they have formed strong root fibres, when they may be planted out. The best method is to plant each cutting in a separate pot of the smallest size. The germinating of the seeds will be greatly promoted by sinking the pots three parts of their depth in a hot bed, keeping them moist and shaded and until they germinate.

Soil, &c. A rich garden mould, composed of light loam, rather sandy than otherwise, with very rotten dung, is desirable for this shrub.

Culture. Most kinds are rapid and luxurious growers, and it is necessary to pay them constant attention in pruning or nipping the extremities of the shoots, or they will soon become ill-formed and straggling. This is particularly requisite during the rains, when heat and moisture combine to increase their growth to excess; allowing them to enjoy the full influence of the sun during the whole of the cold weather, and part of the hot. At the close of the rains, the plants had better be put out into the open ground, and closely pruned, the shoots taken off affording an ample supply of cuttings for multiplying the plants; this putting out will cause them to throw up strong healthy shoots and rich blossoms; but as the hot weather approaches, or in the beginning of March, they must be re-placed in moderate sized pots, with a compost similar to that required for cuttings and placed in the plant shed, as before described. The earth in the pots should be covered with pebbles, or pounded brick of moderate size, which prevents the accumulation of moss or fungi. Geraniums should at no time be over watered, and must at all seasons be allowed a free ventilation.

There is no doubt that if visitors from this to the Cape, would pay a little attention to the subject, the varieties might be greatly increased, and that without much trouble, as many kinds may be produced freely by seed, if brought to the country fresh, and sown immediately on arrival; young plants also in well glazed cases would not take up much space in some of the large vessels coming from thence.

The ANEMONE has numerous varieties, and is, in England, a very favorite flower, but although A. cernua is a native of Japan, and many varieties are indigenous to the Cape, it is very rare here.

The *Double anemone* is the most prized, but there are several *Single* and *Half double* kinds which are very handsome. The stem of a good anemone should be eight or nine inches in height, with a strong upright stalk. The flower ought not to be less than seven inches in circumference, the outer row of petals being well rounded, flat, and expanding at the base, turning up with a full rounded edge, so as to form a well shaped cup, within which, in the double kinds, should arise a large group of long small petals reverted from the centre, and regularly overlapping each other; the colors clear, each shade being distinct in such as are variegated.

The *Garden, or Star Wind flower*, A. hortensis, *Boostan afrooz*, is another variety, found in Persia, and brought thence to Upper India, of a bright scarlet color; a blue variety has also blossomed in Calcutta, and was exhibited at the Show of February, 1847, by Mrs. Macleod, to whom Floriculture is indebted for the introduction of many beautiful exotics heretofore new to India. But it is to be hoped this handsome species of flowering plants will soon be more extensively found under cultivation.

Propagation. Seed can hardly be expected to succeed in this country, as even in Europe it fails of germinating; for if not sown immediately that it is ripe, the length of journey or voyage would inevitably destroy its power of producing. Offsets of the tubers therefore are the only means that are left, and these should not be replanted until they have been a sufficient time out of the ground, say a month or so, to become hardened, nor should they be put into the earth until they have dried, or the whole offset will rot by exposure of the newly fractured side to the moisture of the earth. The tubers should be selected which are plump and firm, as well as of moderate size, the larger ones being generally hollow; these may be obtained in good order from Hobart Town.

Soil, &c. A strong rich loamy soil is preferable, having a considerable portion of well rotted cow-dung, with a little leaf mould, dug to a depth of two feet, and the beds not raised too high, as it is desirable to preserve moisture in the subsoil; if in pots, this is effected by keeping a saucer of water under them continually, the pot must however be deep, or the fibres will have too much wet; an open airy situation is desirable.

Culture. When the plant appears above ground the earth must be pressed well down around the root, as the crowns and tubers are injured by exposure to dry weather, and the plants should be sheltered from the heat of the sun, but not so as to confine the air; they require the morning and evening sun to shine on them, particularly the former.

The IRIS is a handsome plant, attractive alike from the variety and the beauty of its blossoms; some of them are also used medicinally. All varieties produce abundance of seed, in which form the plant might with great care be introduced into this country.

The *Florence Iris*, I. florentina, *Ueersa*, is a large variety, growing some two feet in height, the flower being white, and produced in the hot weather.

The *Persian Iris* I. persica, *Hoobur*, is esteemed not only for its handsome blue and purple flowers, but also for its fragrance, blossoming in the latter part of the cold weather; one variety has blue and yellow blossoms.

The *Chinese Iris*, I. chinensis, *Soosun peelgoosh*, in a small sized variety, but has very pretty blue and purple flowers in the beginning of the hot weather.

Propagation. Besides seed, which should be sown in drills, at the close of the rains, in a sandy soil, it may be produced by offsets.

Soil, &c. Almost any kind of soil suits the Iris, but the best flowers are obtained from a mixture of sandy loam, with leaf mould, the Persian kind requiring a larger proportion of sand.

Culture. Little after culture is required, except keeping the beds clear from weeds, and occasionally loosening the earth. But the roots must be taken, up every two, or at most three years, and replanted, after having been kept to harden for a month or six weeks; the proper season for doing this being when the leaves decay after blossoming.

The TUBEROSE, Polianthes, is well deserving of culture, but it is not by any means a rare plant, and like many indigenous odoriferous flowers, has rather too strong an odour to be borne near at hand, and it is considered unwholesome in a room.

The *Common Tuberose*, P. tuberosa, *Chubugulshubboo*, being a native of India thrives in almost any soil, and requires no cultivation: it is multiplied by dividing the roots. It flowers at all times of the year in bunches of white flowers with long sepals.

The *Double Tuberose*, P. florepleno, is very rich in appearance, and of more delicate fragrance, although still too powerful for the room. Crows are great destroyers of the blossoms, which they appear fond of pecking. This variety is more rare, and the best specimens have been obtained from Hobart Town. It is rather more delicate and requires more attention in culture than the indigenous variety, and should be earthed up, so as to prevent water lodging around the stem.

The LOBELIA is a brilliant class of flowers which may be greatly improved by careful cultivation.

The *Splendid Lobelia*, L. splendens, is found in many gardens, and is a showy scarlet flower, well worthy of culture.

The *Pyramidal Lobelia*, L. pyramidalis, is a native of Nepal, and is a modest pretty flower, of a purple color.

Propagation--is best performed by offsets, suckers, or cuttings, but seeds produce good strong plants, which may with care, be made to improve.

Soil, &c.--A moist, sandy soil is requisite for them, the small varieties especially delighting in wet ground. Some few of this family are annuals, and the roots of no varieties should remain more than three years without renewal, as the blossoms are apt to deteriorate; they all flower during the rains.

The PITCAIRNIA is a very handsome species, having long narrow leaves, with, spined edges and throwing up blossoms in upright spines.

The *Long Stamened Pitcairnia*, P. staminea, is a splendid scarlet flower, lasting long in blossom, which, appears in July or August, and continues till December.

The *Scarlet Pitcairnia*, P. bromeliaefolia, is also a fine rich scarlet flower, but blossoming somewhat sooner, and may be made to continue about a month later.

Propagation--is by dividing the roots, or by suckers, which is best performed at the close of the rains.

Soil, &c. A sandy peat is the favorite soil of this plant, which should be kept very moist.

The DAHLIA, Dahlia; a few years since an attempt was made to rename this beautiful and extensive family and to call it Georgina, but it failed, and it is still better known throughout the world by its old name than the new. It was long supposed that the Dahlia was only found indigenous in Mexico, but Captain Kirke some few years back brought to the notice of the Horticultural Society, that it was to be met with in great abundance in Dheyra Dhoon, producing many varieties both single and double; and he has from time to time sent down quantities of seed, which have greatly assisted its increase in all parts of India. It has also been found in Nagpore.

A good Dahlia is judged of by its form, size, and color. In respect to the first of these its *form* should be perfectly round, without any inequalities of projecting points of the petals, or being notched, or irregular. These should also be so far revolute that the side view should exhibit a perfect semicircle in its outline, and the eye or prolific disc, in the centre should be entirely concealed. There has been recently introduced into this country a new variety, all the petals of which are quilled, which has a very handsome appearance.

In *size* although of small estimation if the other qualities are defective, it is yet of some consideration, but the larger flowers are apt to be wanting in that perfect hemispherical form that is so much admired.

The *color* is of great importance to the perfection of the flower; of those that are of one color this should be clear, unbroken, and distinct; but when mixed hues are sought, each color should be clearly and distinctly defined without any mingling of shades, or running into each other. Further, the flowers ought to be erect so as to exhibit the blossom in the fullest manner to the view. The most usual colors of the imported double Dahlias, met with in India, are crimson, scarlet, orange, purple, and white. Amongst those raised from seed from. Dheyra Dhoon[137] of the double kind, there are of single colors, crimson, deep crimson approaching to maroon, deep lilac, pale lilac, violet, pink, light purple, canary color, yellow, red, and white; and of mixed colors, white and pink, red and yellow, and orange and white: the single ones of good star shaped flowers and even petals being of crimson, puce, lilac, pale lilac, white, and orange. Those from Nagpore seed have yielded, double flowers of deep crimson, lilac, and pale purple, amongst single colors; lilac and blue, and red and yellow of mixed shades; and single flowered, crimson, and orange, with mixed colors of lilac and yellow, and lilac and white.

Propagation--is by dividing the roots, by cuttings, by suckers, or by seed; the latter is generally resorted to, where new varieties are desired. Mr. George A. Lake, in an article on this subject (*Gardeners' Magazine*, 1833) says: "I speak advisedly, and from, experience, when I assert that plants raised from cuttings do not produce equally perfect flowers, in regard to size, form, and fulness, with those produced by plants grown from division of tubers;" and he more fully shews in another part of the same paper, that this appears altogether conformable to reason, as the cutting must necessarily for a long period want that store of starch, which is heaped up in the full grown tuber for the nutriment of the plant. This objection however might be met by not allowing the cuttings to flower in the season when they are struck.

To those who are curious in the cultivation of this handsome species, it may be well to know how to secure varieties, especially of mixed colors; for this purpose it is necessary to cover the blossoms intended for fecundation with fine gauze tied firmly to the foot stalk, and when it expands take the pollen from the male flowers with a camel's hair pencil, and touch with it each floret of the intended bearing flower, tying the gauze again over it, and keeping it on until the petals are withered. The operation requires to be performed two or three successive days, as the florets do not expand together.

Soil &c. They thrive best in a rich loam, mixed with sand; but should not be repeated too often on the same spot, as they exhaust the soil considerably.

Culture. The Dahlia requires an open, airy position unsheltered by trees or walls, the plants should be put out where they are to blossom, immediately on the cessation of the rains, at a distance of three feet apart, either in rows or in clumps, as they make a handsome show in a mass; and as they grow should be trimmed from the lower shoots, to about a foot in height, and either tied carefully to a stake, or, what is better, surrounded by a square or circular trellis, about five feet in height. As the buds form they should be trimmed off, so as to leave but one on each stalk, this being the only method by which full, large, and perfectly shaped blossoms are obtained. Some people take up the tubers every year in February or March, but this is unnecessary. The plants blossom in November and December in the greatest perfection, but may with attention be continued from the beginning of October to the end of February.

Those plants which are left in the ground during the whole year should have their roots opened immediately on the close of the rains, the superabundant or decayed tubers, and all suckers being removed, and fresh earth filled in. The earth should always be heaped up high around the stems, and it is a good plan to surround each plant with a small trench to be filled daily with water so as to keep the stem and leaves dry.

The PINK, Dianthus, *Kurunful*, is a well known species of great variety, and acknowledged beauty.

The *Carnation*, D. caryophyilus, *Gul kurunful*, is by this time naturalized in India, adding both beauty and fragrance to the parterre; the only variety however that has yet appeared in the country is the clove, or deep crimson colored: but the success attending the culture of this beautiful flower is surely an encouragement to the introduction of other sorts, there being above four hundred kinds, especially as they may be obtained from seed or pipings sent packed in moss, which will remain in good condition for two or three months, provided no moisture beyond what is natural to the moss, have access to them.

The distinguishing marks of a good carnation may be thus described: the stem should be tall and straight, strong, elastic, and having rather short foot stalks, the flower should be fully three inches in diameter with large well formed petals, round and uncut, long and broad, so as to stand out well, rising about half an inch above the calyx, and then the outer ones turned off in a horizontal direction, supporting those of the centre, decreasing gradually in size, the whole forming a near approach to a hemisphere. It flowers in April and May.

Propagation--is performed either by seed, by layers, or by pipings; the best time for making the two latter is when the plant is in full blossom, as they then root more strongly. In this operation the lower leaves should be trimmed off, and an incision made with a sharp knife, by entering the knife about a quarter of an inch below the joint, passing it through its centre; it must then be pegged down with a hooked peg, and covered with about a quarter of an inch of light rich mould; if kept regularly moist, the layers will root in about a month's time: they may then be taken off and planted out into pots in a sheltered situation, neither exposed to excessive rain, nor sun, until they shoot out freely.

Pipings (or cuttings as they are called in other plants) must be taken off from a healthy, free growing plant, and should have two complete joints, being cut off horizontally close under the second one; the extremities of the leaves must also be shortened, leaving the whole length of each piping two inches; they should be thrown into a basin of soft water for a few minutes to plump them, and then planted out in moist rich mould, not more than an inch being inserted therein, and slightly watered to settle the earth close around them; after this the soil should be kept moderately moist, and never exposed to the sun. Seed is seldom resorted to except to introduce new varieties.

Soil, &c.--A mixture of old well rotted stable manure, with one-third the quantity of good fine loamy earth, and a small portion of sand, is the best soil for carnations.

Culture.--The plants should be sheltered from too heavy a fall of rain, although they require to be kept moderately moist, and desire an airy situation. When the flower stalks are about six or eight inches in height, they must be supported by sticks, and, if large full blossoms be sought for, all the buds, except the leading one, must be removed with a pair of scissors; the calyx must also be frequently examined, as it is apt to burst, and if any disposition to this should appear, it will be well to assist the uniform expansion by cutting the angles with a sharp penknife. If, despite all precautions the calyx burst and let out the petals, it should be carefully tied with thread, or a circular piece of card having a hole in the centre should be drawn over the bud so as to hold the petals together, and display them to advantage by the contrast of the white color.

Insects, &c.--The most destructive are the red, and the large black ant, which attack, and frequently entirely destroy the roots before you can be aware of its approach; powdered turmeric should therefore be constantly kept strewed around this flower.

The *Common Pink*, Dianthus Chinensis, *Kurunful*, and the *Sweet William*, D: barbatus, are pretty, ornamental plants, and may be propagated and cultivated in the same way as the carnation, save that they do not require so much care, or so good a soil, any garden mould sufficing; they are also more easily produced from seed.

The VIOLET, Viola, *Puroos*, is a class containing many beautiful flowers, some highly ornamental and others odoriferous.

The *Sweet Violet*, V. odorata, *Bunufsh'eh*, truly the poet's flower. It is a deserved favorite for its delightful fragrance as well as its delicate and retiring purple flowers; there is also a white variety, but it is rare in this country, as is also the double kind. This blossoms in the latter part of the cold weather.

The *Shrubby Violet*, V. arborescens, or suffruticosa, *Rutunpuroos*, grows wild in the hills, and is a pretty blue flower, but wants the fragrance of the foregoing.

The *Dog's Violet*, V. canina, is also indigenous in the hills.

Propagation.--All varieties may be propagated by seed, but the most usual method is by dividing the roots, or taking off the runners.

Soil, &c.--The natural *habitat* of the indigenous varieties is the sides and interstices of the rocks, where leaf mould, and micaceous sand, has accumulated and moisture been retained, indicating that the kind of soil favorable to the growth of this interesting little plant is a rich vegetable mould, with an admixture of sand, somewhat moist, but having a dry subsoil.

Culture.--It would not be safe to trust this plant in the open ground except during a very short period of the early part of the cold weather, when the so doing will give it strength to form blossoms. In January, however, it should be re-potted, filling the pots about half-full of pebbles or stone-mason's cuttings, over which should be placed good rich vegetable mould, mixed with a large proportion of sand, covering with a thin layer of the same material as has been put into the bottom of the pot; a top dressing of ground bones is said to improve the fineness of the blossoms. They should not be kept too dry, but at the same time watered cautiously, as too much of either heat or moisture destroys the plants.

The *Pansy* or *Heart's-ease*, V. tricolor, *Kheeroo, kheearee*, derives its first name from the French *Pensée*. It was known amongst the early Christians by the name of *Flos Trinitatis*, and worn as a symbol of their faith. The high estimation which it has of late years attained in Great Britain as a florist's flower has, in the last two or three years, extended itself to this country. There are nearly four hundred varieties, a few of which only have been found here.

The characters of a fine Heart's-ease are, the flower being well expanded, offering a flat, or if any thing, rather a revolute surface, and the petals so overlapping each other as to form a circle without any break in the outline. These should be as nearly as possible of a size, and the greater length of the two upper ones concealed by the covering of those at the side in such manner as to preserve the appearance of just proportion: the bottom petal being broad and two-lobed, and well expanded, not curving inwards. The eye should be of moderate, or rather small size, and much additional beauty is afforded, if the pencilling is so arranged as to give the appearance of a dark angular spot. The colors must also be clear, bright, and even, not clouded or indistinct. Undoubtedly the handsomest kinds are those in which the two upper petals are of deep purple and the triade of a shade less: in all, the flower stalk should be long and stiff. The plant blossoms in this country in February and March, although it is elsewhere a summer flower.

Propagation.--In England the moat usual methods are dividing the roots, layers, or cuttings from the stem, and these are certainly the only sure means of preserving a good variety; but it is almost impossible in India to preserve the plant through the hot weather, and therefore it is more generally treated as an annual, and raised every year from seed, which should be sown at the close of the rains; as however their growth, in India is as yet little known, most people put the imported seed into pots as soon as it arrives, lest the climate should deteriorate its germinating power, as it is well known, that even in Europe the seed should be sown as soon as possible after ripening. It will be well also to assist its sprouting with a little bottom heat, by plunging the pot up to its rim in a hot bed. American seed should be avoided as the blossoms are little to be depended on, and generally yield small, ill-formed flowers, clouded and run in color.

Soil, &c.--This should be moist, and the best compost is formed of one-sixth of well rotted dung from an old hot bed, and five-sixth of loam, or one-fourth of leaf mould and the remainder loam, but in either case well incorporated and exposed for some time previous to use to the action of the sun and air by frequent turning.

Culture.--A shady situation is to be preferred, especially for the dark varieties which assume a deeper hue if so placed. But it has been observed by Mackintosh, that "the light varieties bloomed lighter in the shade, and darker in the sunshine--a very remarkable effect, for which I cannot account." The plants must at all times be kept moist, never being allowed to become dry, and should be so placed as to receive only the morning sun before ten o'clock. Under good management the plants will extend a foot or more in height, and have a handsome appearance if trained over a circular trellis of rattan twisted. When they rise too high, or it is desirable to fill out with side shoots, the tops must be pinched off, and larger flowers will be obtained if the flower buds are thinned out where they appear crowded.

These plants look very handsome when grown in large masses of several varieties, but the seeds of those grown in this manner should not be made use of, as they are sure to sport; to prevent which it is also necessary that the plants which it is desired to perpetuate in this manner should be isolated at a distance from any other kind, and it would be advisable to cover them with thin gauze to prevent impregnation from others by means of the bees and other insects. For show flowers the branches should be kept down, and not suffered to straggle out or multiply; these will also be improved by pegging the longer branches down under the soil, and thereby increasing the number of the root fibres, hence adding to their power of accumulating nourishment, and not allowing them to expand beyond a limited number of blossoms, and those retained should be as nearly equal in age as possible.

The HYDRANGEA is a hardy plant requiring a good deal of moisture, being by nature an inhabitant of the marshes.

The *Changeable Hydrangea*, H. hortensis, is of Chinese origin and a pretty growing plant that deserves to be a favorite; it blossoms in bunches of flowers at the extremities of the branches which are naturally pink, but in old peat earth, or having a mixture of alum, or iron filings, the color changes to blue. It blooms in March and April.

Propagation may be effected by cuttings, which root freely, or by layers.

Soil, &c.--Loam and old leaf mould, or peat with a very small admixture of sand suits this plant. Their growth is much promoted by being turned out, for a month or two in the rains, into the open ground, and then re-potted with new soil, the old being entirely removed from the roots: and to make it flower well it must not be encumbered with too many branches.

The HOYA is properly a trailing plant, rooting at the joints, but have been generally cultivated here as a twiner.

The *Fleshy-leaved Hoya*, H. carnosa, is vulgarly called the wax flower from its singular star shaped-whitish pink blossoms, with a deep colored varnished centre, having more the appearance of a wax model than a production of nature. The flowers appear in globular groups and have a very handsome appearance from the beginning of April to the close of the rains.

The *Green flowered Hoya*, H. viridiflora, *Nukchukoree, teel kunga*, with its green flowers in numerous groups, is also an interesting plant, it is esteemed also for its medicinal properties.

Propagation.--Every morsel of these plants, even a piece of the leaf, will form roots if put in the ground, cuttings therefore strike very freely, as do layers, the joints naturally throwing out root-fibres although not in the earth.

Soil, &c.--A light loam moderately dry is the best for these plants, which look well if trained round a circular trellis in the open border.

The STAPELIA is an extensive genus of low succulent plants without leaves, but yielding singularly handsome star-shaped flowers; they are of African origin growing in the sandy deserts, but in a natural state very diminutive being increased to their present condition and numerous varieties by cultivation, they mostly have an offensive smell whence some people call them the carrion plant. They deserve more attention than has hitherto been shown to them in India.

The *Variegated Stapelia*, S. variegata, yields a flower in November, the thick petals of which are yellowish green with brown irregular spots, it is the simplest of the family.

The *Revolute-flowered Stapelia*, S. revoluta, has a green blossom very fully sprinkled with deep purple, it flowers at the close of the rains.

The *Toad Stapelia*, S. bufonia, as its name implies, is marked like the back of the reptile from whence it has its name; it flowers in December and January.

The *Hairy Stapelia*, S. hirsuta, is a very handsome variety, being, like the rest, of green and brown, but the entire flower covered with fine filaments or hairs of a light purple, at various periods of the year.

The *Starry Stapelia*, S. stellaris, is perhaps the most beautiful of the whole, being like the last covered with hairs, but they are of a bright pinkish blue color; there appears to be no fixed period for flowering.

The HAIRY CARRULLUMA, C. crinalata, belongs to the same family as the foregoing species, which it much resembles, except that it blossoms in good sized globular groups of small star-shaped flowers of green, studded and streaked with brown.

Propagation is exceedingly easy with each of the last named two species; as the smallest piece put in any soil that is moist, without being saturated, will throw out root fibres.

Soil, &c.--This should consist of one-half sand, one-fourth garden mould, and one-fourth well rotted stable manure. The pots in which they are planted should have on the top a layer of pebbles, or broken brick. All the after culture they require is to keep them within bounds, removing decayed portions as they appear and avoiding their having too much moisture.

The perennial border plants, besides those included above, are very numerous; the directions for cultivation admitting, from their similarity, of the following general rules:--

Propagation.--Although some few will admit of other modes of multiplication, the most usually successful are by seed, by suckers, or by offsets, and by division of the root, the last being applicable to nine-tenths of the hardy herbaceous plants, and performed either by taking up the whole plant and gently separating it by the hand, or by opening the ground near the one to be divided, and cutting off a part of the roots and crown to make new the sections being either at once planted where they are to stand, or placed for a short period in a nursery; the best time for this operation is the beginning of the rains. Offsets or suckers being rapidly produced during the rains, will be best removed towards their close, at which period, also, seed should be sown to benefit by the moisture remaining in the soil. The depth at which seeds are buried in the earth varies with their magnitude, all the pea or vetch kind will bear being put at a depth of from half an inch to one inch; but with the smallest seeds it will be sufficient to scatter them, on the sifted soil, beating them down with, the palm of the hand.

Culture.--Transplanting this description of plants will be performed to best advantage during the rains. The general management is comprehended in stirring the soil occasionally in the immediate vicinity of the roots; taking up overgrown plants, reducing and replanting them, for which the rains is the best time; renewing the soil around the roots; sticking the weak plants; pruning and trimming others, so as to remove all weakly or decayed parts.

Once a year, before the rains, the whole border should be dug one or two spits deep, adding soil from the bottom of a tank or river; and again, in the cold weather, giving a moderate supply of well rotted stable manure, and leaf mould in equal portions.

Crossing is considered as yet in its infancy even in England, and has, except with the Marvel of Peru, hardly even been attempted in this country. The principles under which this is effected are fully explained at page 27 of the former part of this work; but it may also be done in the more woody kinds by grafting one or more of the same genus on the stock of another, the seed of which would give a new variety.

Saving seed requires great attention in India, as it should be taken during the hot weather if possible; to effect which the earliest blossoms must be preserved for this purpose. With some kinds it will be advisable to assist nature by artificial impregnation with a camel hair pencil, carefully placing the pollen on the point of the stigma. The seeds should be carefully dried in some open, airy place, but not exposed to the sun, care being afterwards taken that they shall be deposited in a dry place, not close or damp, whence the usual plan of storing the seeds in bottles is not advisable.

BULBS.

Bulbs have not as yet received that degree of attention in this country (India) that they deserve, and they may be considered to form a separate class, requiring a mode of culture differing from that of others. Their slow progress has discouraged many and a supposition that they will only thrive in the Upper Provinces, has deterred others from attempting to grow them, an idea which has also been somewhat fostered by the Horticultural Society, when they received a supply from England, having sent the larger portion of them to their subscribers in the North West Provinces.

The NARCISSUS will thrive with care, in all parts of India, and it is a matter of surprise that it is not more frequently met with. A good Narcissus should have the six petals well formed, regularly and evenly disposed, with a cup of good form, the colors distinct and clear, raised on strong erect stems, and flowering together.

The *Polyanthes Narcissus*, N. tazetta, *Narjus, hur'huft nusreen*, is of two classes, white and sulphur colored, but these have sported into almost endless varieties, especially amongst the Dutch, with whom this and most other bulbs are great favorites. It flowers in February and March.

The *Poet's Narcissus*, N. poeticus, *Moozhan, zureenkuda* is the favorite, alike for its fragrance and its delicate and graceful appearance, the petals being white and the cup a deep yellow: it flowers from the beginning of January to the end of March and thrives well. The first within the recollection of the author, in Bengal, was at Patna, nearly twelve years since, in possession of a lady there under whose care it blossomed freely in the shade, in the month of February.

The *Daffodil*, N. pseudo-narcissus, *Khumsee buroonk*, is of pale yellow, and some of the double varieties are very handsome.

Propagation is by offsets, pulled off after the bulbs are taken out of the ground, and sufficiently hardened.

Soil, &c.--The best is a fresh, light loam with some well rotted cow dung for the root fibres to strike into, and the bottom of the pot to the height of one-third filled with pebbles or broken brick. They will not blossom until the fifth year, and to secure strong flowers the bulbs should only be taken up every third year. An eastern aspect where they get only the morning sun, is to be preferred. The PANCRATIUM is a handsome species that thrives well, some varieties being indigenous, and others fully acclimated, generally flowering about May or June.

The *One-flowered Pancratium*, P. zeylanicum, is rather later than the rest in flowering and bears a curiously formed white flower.

The *Two-flowered Pancratium*, P. triflorum, *Sada kunool*, was so named by Roxburg, and gives a white flower in groups of threes, as its name implies.

The *Oval leaved pancratium*, P. ovatum, although of West Indian origin, is so thoroughly acclimated as to be quite common in the Indian Garden.

Propagation.--The best method is by suckers or offsets which are thrown out very freely by all the varieties.

Soil, &c.--Any common garden soil will suit this plant, but they thrive best with a good admixture of rich vegetable mould.

The HYACINTH, Hyacinthus, is an elegant flower, especially the double kind. The first bloomed in Calcutta was exhibited at the flower show some three years since, but proved an imperfect blossom and not clear colored; a very handsome one, however, was shown by Mrs. Macleod in February 1847, and was raised from a stock originally obtained at Simlah. The Dutch florists have nearly two thousand varieties.

The distinguishing marks of a good hyacinth are clear bright colors, free from clouding or sporting, broad bold petals, full, large and perfectly doubled, sufficiently revolute to give the whole mass a degree of convexity: the stem strong and erect and the foot stalks horizontal at the base, gradually taking an angle upwards as they approach the crown, so as to place the flowers in a pyramidical form, occupying about one-half the length of the stem.

The *Amethyst colored Hyacinth*, H. amethystimus, is a fine handsome flower, varying in shade from pale blue to purple, and having bell shaped flowers, but the foot stalks are generally not strong and they are apt to become pendulous.

The *Garden Hyacinth*, H. orientalis, *Sumbul, abroad*, is the handsomer variety, the flowers being trumpet shaped, very double and of varying colors--pink, red, blue, white, or yellow, and originally of eastern growth. It flowers in February and has considerable fragrance.

Propagation.--In Europe this is sometimes performed by seed, but as this requires to be put into the ground as soon as possible after ripening, and moreover takes a long time to germinate, this method would hardly answer in this country, which must therefore, at least for the present, depend upon imported bulbs and offsets.

Soil, &c.--This, as well as its after culture, is the same as for the Narcissus. They will not show flowers until the second year, and not in good bloom before the fifth or sixth of their planting out.

The CROCUS, Crocus lutens, having no native name, has yet, it is believed, been hardly ever known to flower here, even with the utmost care. A good crocus has its colors clear, brilliant, and distinctly marked.

Propagation--must be effected, for new varieties, by seeds, but the species are multiplied by offsets of the bulb.

Soil, &c. Any fair garden soil is good for the crocus, but it prefers that which is somewhat sandy.

Culture. The small bulbs should be planted in clumps at the depth of two inches; the leaves should not be cut off after the plant has done blossoming, as the nourishment for the future season's flower is gathered by them.

The IXIA, is originally from the Cape, and belongs to the class of Iridae: the Ixia Chinensis, more properly Morea Chinensis, is a native of India and China, and common in most gardens.

Propagation--is by offsets.

Soil, &c. The best is of peat and sand, it thrives however in good garden soil, if not too stiff, and requires no particular cultivation.

The LILY, Lilium, *Soosun*, the latter derived from the Hebrew, is a handsome species that deserves more care than it has yet received in India, where some of the varieties are indigenous.

The *Japan Lily*, L. japonicum, is a very tall growing plant, reaching about 5 feet in height with broad handsome flowers of pure white, and a small streak of blue, in the rains.

The *Daunan Lily*, L. dauricum, *Rufeef, soosun*, gives an erect, light orange flower in the rains.

The *Canadian lily*, L. Canadense B'*uhmutan*, flowers in the rains in pairs of drooping reflexed blossoms of a rather darker orange, sometimes spotted with a deeper shade.

Propagation--is effected by offsets, which however will not flower until the third or fourth year.

Soil, &c. This is the same as for the Narcissus, but they do not require taking up more frequently than once in three years, and that only for about a month at the close of the rains, the Japan lily will thrive even under the shade of trees.

The AMARYLLIS is a very handsome flower, which has been found to thrive well in this country, and has a great variety, all of which possess much beauty, some kinds are very hardy, and will grow freely in the open ground.

The *Mexican Lily*, A. regina Mexicanae, is a common hardy variety found in most gardens, yielding an orange red flower in the months of March and April, and will thrive even under the shades of trees.

The *Ceylonese Amaryllis*, A: zeylanica, *Suk'h dursun*, gives a pretty flower about the same period.

The *Jacoboean Lily*, A, formosissima, has a handsome dark red flower of singular form, having three petals well expanded above, and three others downwards rolled over the fructile organs on the base, so as to give the idea of its being the model whence the Bourbon *fleur de lis* was taken, the stem is shorter than the two previous kinds, blossoming in April or May.

The *Noble Amaryllis*, A: insignia, is a tall variety, having pink flowers in March or April.

The *Broad-leaved Amaryllis*, A: latifolia, is a native of India with pinkish white flowers about the same period of the year.

The *Belladonna Lily*. A: belladonna is of moderately high stem, supporting a pink flower of the same singular form as the Jacoboean lily, in May and June.

Propagation--is by offsets of the bulb, which most kinds throw out very freely, sometimes to the extent of ten, or a dozen in the season.

Soil, &c.--For the choice kinds is the same as is required for the narcissus, and water should on no account be given over the leaves or upper part of the bulb.

The common kinds look well in masses, and a good form of planting them is in a series of raised circles, so as for the whole to form a round bed.

The DOG'S TOOTH VIOLET, Erythronium, is a pretty flowering bulb and a great favorite with florists in Europe.

The *Common Dog's tooth Violet*, E. dens canis, is ordinarily found of reddish purple, there is also a white variety, but it is rare, neither of them grow above three or four inches in height, and flower in March or April.

The *Indian Dog's tooth Violet*, E. indicum, *junglee kanda*, is found in the hills, and flowers at about the same time, with a pink blossom.

The SUPERB GLORIOSA, Gloriosa superba, *Kareearee, eeskooee langula*, is a very beautiful species of climbing bulb, a native of this country, and on that account neglected, although highly esteemed as a stove plant in England; the leaves bear tendrils at the points, and the flower, which is pendulous, when first expanded, throws its petals nearly erect of yellowish green, which gradually changes to yellow at the base and bright scarlet at the point; the pistil which shoots from the seed vessel horizontally possesses the singular property of making an entire circuit between sun-rise and sun-set each day that the flower continues, which is generally for some time, receiving impregnation from every author as it visits them in succession. It blooms in the latter part of the rains.

Propagation is in India sometimes from seed, but in Europe it is confined to division of the offsets.

Soil, &c.--Most garden soils will suit this plant, but it affords the handsomest, and richest colored flowers in fresh loam mixed with peat or leaf mould, without dung. It should not have too much water when first commencing its growth, and it requires the support of a trellis over which it will bear training to a considerable extent, growing to the height of from five to six feet.

MANY OTHER BULBS, there is no doubt, might be successfully grown in India where every thing is favorable to their growth, and so much facility presents itself for procuring them from the Cape of Good Hope; the natural *habitat* of so many varieties of the handsomest species, nearly all of them flowering between the end of the cold weather and the close of the rains.

Some of these being hardy, thrive in the open ground with but little care or trouble, others requiring very great attention, protection from exposure, and shelter from the heat of the sun, and the intensity of its rays; which should therefore have a particular portion of the plant-shed assigned to them, such being inhabitants of the green house in colder climates, and the reason of assigning them such separated part of the chief house, or what is better perhaps, a small house to themselves, is that in culture, treatment, and other respects they do not associate with plants of a different character.

One great obstacle which the more extensive culture of bulbs has had to contend against, may be found in that impatience that refuses to give attention to what requires from three to five years to perfect, generally speaking people in India prefer therefore to cultivate such plants only as afford an immediate result, especially with relation to the ornamental classes.

Propagation.--The bulb after the formation of the first floral core is instigated by nature to continue its species, as immediately the flower fades the portion of bulb that gave it birth dies, for which purpose it each year forms embryo bulbs on each side of the blossoming one, and which although continued in the same external coat, are each perfect and complete plants in themselves, rising from the crown of the root fibres: in some kinds this is more distinctly exhibited by being as it were, altogether outside and distinct from, the main, or original bulb. These being separated for what are called offsets, and should be taken off only when the parent bulb has been taken up and hardened, or the young plant will suffer.

Some species of bulbous rooted plants produce seeds, but this method of reproduction, can seldom be resorted to in this country, and certainly not to obtain new kinds, as the seeds require to be sown as soon as ripe.

Soil, Culture, &c.--For the delicate and rare bulbs, it is advisable to have pots purposely made of some fifteen inches in height with a diameter of about seven or eight inches at the top, tapering down to five, with a hole at the bottom as in ordinary flower pots, and for this to stand in, another pot should be made without any hole, of a height of about four inches, sufficient size to leave the space of about an inch all round between the outer side of the plant pot and the inner side of the smaller pot or saucer.

This will allow the plant pot to be filled with crocks, pebbles, or stone chippings to the height of five inches, or about an inch higher than the level of the water in the saucer, above which may be placed eight inches in depth of soil and one inch on the top of that, pebbles or small broken brick. By this arrangement, the saucer being kept filled, or partly filled, as the plant may require, with water, the fibres of the root obtain a sufficiency of moisture for the maintenance and advancement of the plant without chance of injury to the bulb or stem, by applying water to the upper earth which is also in this prevented from becoming too much saturated. Light rich sandy loam, with a portion of sufficiently decomposed leaf mould, is the best soil for the early stages of growing bulbs.

So soon as the leaves change color and wither, then all moisture must be withheld, but as the repose obtained by this means is not sufficient to secure health to the plant, and ensure its giving strong blossoms, something more is required to effect this purpose. This being rendered the more necessary because in those that form offsets by the sides of the old bulbs, they would otherwise become crowded and degenerate, the same occurring also with those forming under the old ones, which will get down so deep that they cease to appear.

The time to take up the bulb is when the flower-stem and leaves have commenced decay; taking dry weather for the purpose, if the bulbs are hardy, or if in pots having reduced the moisture as above shown, but it must be left to individual experience to discover how long the different varieties should remain out of the ground, some requiring one month's rest, and others enduring three or four, with advantage; more than that is likely to be injurious. When out of the ground, during the first part of the period they are so kept, it should be, say for a fortnight at least, in any room where no glare exists, with free circulation of air, after which the off-sets may be removed, and the whole exposed to dry on a table in the verandah, or any other place that is open to the air, but protected from the sunshine, which would destroy them.

Little peculiarity of after treatment is requisite, except perhaps that the bulbs which are to flower in the season should have a rather larger proportion of leaf mould in the compost, and that if handsome flowers are required, it will be well to examine the bulb every week at least by gently taking the mould from around them, and removing all off-sets that appear on the old bulb. For the securing strength to the plant also, it will be well to pinch off the flower so soon as it shews symptoms of decay.

The wire worm is a great enemy to bulbs, and whenever it appears they should be taken up, cleaned, and re-planted. It is hardly necessary to say that all other vermin and insects must be watched, and immediately removed.

THE BIENNIAL BORDER PLANTS.

It is only necessary to mention a few of these, as the curious in floriculture will always make their own selection, the following will therefore suffice.--

The SPEEDWELL-LEAVED HEDGE HYSSOP, Gratiola veronicifolia, *Bhoomee, sooél chumnee*, seldom cultivated, though deserving to be so, has a small blue flower.

The SIMPLE-STALKED LOBELIA, Lobelia simplex, introduced from the Cape, yields a pretty blue flower.

The EVENING PRIMROSE, Oenothera mutabilis, a pretty white flower that blossoms in the evening, its petals becoming pink by morning.

The FLAX-LEAVED PIMPERNEL, Anagallis linifolia, a rare plant, giving a blue flower in the rains; introduced from Portugal.

The BROWALLIA, of two lauds, both pretty and interesting plants; originally from South America.

The *Spreading Browallia*, B. demissa is the smallest of these, and blossoms in single flowers of bright blue, at the beginning of the cold weather.

The *Upright Browallia*, B. alata, gives bloom in groups, of a bright blue; there is also a white variety, both growing to the height of nearly two feet.

The SMALL-FLOWERED TURNSOLE, Heliotropium parviflorum, *B'hoo roodee*, differs from the rest of this family which are mostly perennials; it yields groups of white flowers, which are fragrant.

The FLAX-LEAVED CANDYTUFT, Iberis linifolia, with its purple blossoms, is very rare, but it has been sometimes grown with, success.

The STOCK, Mathiola, is a very popular plant, and deserves more extensive cultivation in this country.

The *Great Sea Stock*, M sinuata, is rare and somewhat difficult to bring into bloom, it possesses some fragrance and its violet colored groups of flowers have rather a handsome appearance about May.

The *Ten weeks' Stock*, M annua, is also a pleasing flower about the same time. In England this is an annual, but here it is not found to bloom freely until the second year, its color is scarlet, and it has some fragrance.

The *Purple Gilly flower*, M incana, is a pretty flower of purple color, and fragrant. There are some varieties of it such as the *Double*, multiplex, the *Brompton*, coccinea, and the *White*, alba, varying in color and blossoming in April.

The STARWORT, Aster, is a hardy flowering plant not very attractive, except as it yields blossoms at all seasons, if the foot stalks are cut off as soon as the flower has faded, there are very numerous varieties of this plant which is, in Europe a perennial, but it is preferable to treat it here as only biennial, otherwise it degenerates.

The *Bushy Starwort*, A dumosus, is a free blossoming plant in the rains, with white flowers.

The *Silky leaved Starwort*, A. sericeus, is Indigenous in the hills, putting forth its blue blossoms during the rains.

The *Hairy Starwort*, A pilosus, is of very pale blue, and may, with care, be made to blossom throughout the year.

The *Chinese Starwort*, A chinensis, is of dark purple and very prolific of blossoms at all times.

The BEAUTIFUL JUSTICIA, J speciosa, although, described by Roxburgh as a perennial, degenerates very much after the second year, it affords bright carmine colored flowers at the end of the cold weather.

The COMMON MARVEL OF PERU, Mirabilis Jalapa *Gul abas, krushna kelee,* is vulgarly called the Four o'clock from its blossoms expanding in the afternoon. There are several varieties distinguished only by difference of color, lilac, red, yellow, orange, and white, which hybridize naturally, and may easily be obliged to do so artificially, if any particular shades are desired.

The HAIRY INDIGO, Indigofera hirsuta, yields an ornamental flower with abundance of purple blossoms.

The HIBISCUS This class numbers many ornamental plants, the blossoms of which all maintain the same character of having a darkened spot at the base of each petal.

The *Althaea frutex*, H syriacus, *Gurhul,* yields a handsome purple flower in the latter part of the rains, there are also a white, and a red variety.

The *Stinging Hibiscus* H pruriens, has a yellow flower at the same season.

The *Hemp leaved Hibiscus*, H cannabinus, *Anbaree*, is much the same as the last.

The *Bladder Ketmia*, H trionum, is a dwarf species, yellow, with a brown spot at the base of the petal.

The *African Hibiscus* H africanus, is a very handsome flower growing to a considerable height, expanding to the diameter of six to seven inches, of a bright canary color, the dark blown spots at the base of the petals very distinctly marked, the seeds were considered a great acquisition when first obtained from Hobarton, but the plant has since been seen in great perfection growing wild in the *Turaee* at the foot of the Darjeeling range of hills, blooming in great perfection at the close of the rains.

The *Chinese Hibiscus*, H rosa sinensis, *Jooua, jasoon, jupa*, although, really a perennial flower, is in greatest perfection if kept as a biennial, it flowers during the greater part of the season a dark red flower with a darker hued spot, there are also some other varieties of different colors yellow, scarlet, and purple.

The TREE MALLOW, Lavatera arborea, has of late years been introduced from Europe, and may now be found in many gardens in India yielding handsome purple flowers in the latter part of the rains.

But it is unnecessary to continue such a mere catalogue, the character and general cultivation of which require no distinct rules, but may all be resolved into one general method, of which the following is a sketch.

Propagation--They are all raised from seed, but the finest double varieties require to be continued by cuttings. The seed should be sown as soon as it can after opening, but if this occur during the rains, the beds, or pots, perhaps better, must be sheltered, removing the plants when they are few inches high to the spot where they are to remain, care being at the same time taken in removing those that have tap roots, such as Hollyhock, Lavatera, &c not to injure them, as it will check their flowering strongly, the best mode is to sow those in pots and transplant them, with balls of earth entire, into the borders, at the close of the rains. Cuttings of such as are multiplied by that method, are taken either from the flower stalks, or root-shoots, early in the rains, and rooted either in pots, under shelter, or in beds, protected from the heavy showers.

Culture--Cultivation after the plants are put into the borders, is the same as for perennial plants. But the duration and beauty of the flowers is greatly improved by cutting off the buds that shew the earliest, so as to retard the bloom--and for the same reason the footstalk should be cut off when the flowers fade, for as soon as the plant begins to form seed, the blossoms deteriorate.

THE ANNUAL BORDER PLANTS.

These are generally known to every one, and many of them are so common as hardly to need notice, a few of the most usual are however mentioned, rather to recal the scattered thoughts of the many, than as a list of annuals.

The MIGNIONETTE, Resoda odorata, is too great a favorite both on account of its fragrance and delicate flowers not to be well known, and by repeated sowings it may be made under care to give flowers throughout the year but it is advisable to renew the seed occasionally by fresh importations from Europe, the Cape, or Hobarton.

The PROLIFIC PINK, Dianthus prolifer *Kurumful,* is a pretty variety; that blossoms freely throughout the year, sowing to keep up succession, the shades and net work marks on them are much varied, and they make a very pretty group together.

The LUPINE, Lupinus, is a very handsome class of annuals, many of which grow well in India, all of them flowering in the cold season.

The *Small blue Lupine,* L. varius, was introduced from the Cape and is the only one noticed by Roxburgh.

The *Rose, and great blue Lupine,* L. pilosus and hirsutus, are both good sized handsome flowers.

The *Egyptian, or African Lupins,* L. thermis, *Turmus,* is the only one named in the native language, and has a white flower.

The *Tree Lupine,* L. arboreus, is a shrubby plant with a profusion of yellow flowers which has been successfully cultivated from Hobarton seed.

The CATCHFLY, Silene, the only one known here is the small red, S. rubella, having a very pretty pink flower appearing in the cold weather.

The LARKSPUR, Delphinum, has not yet received any native name, and deserves to be much more extensively cultivated, especially the Neapolitan and variegated sorts. The common purple, D. Bhinensis, being the one usually met with; it should be sown in succession from September to December, but the rarer kinds must not be put in sooner than the middle of November, as these do not blossom well before February, March, or April.

The SWEET PEA, Lathyrus odoralus, is not usually cultivated with success, because it has been generally sown too late in the season, to give a sufficient advance to secure blossoming. The seeds should be put in about the middle of the rains in pots and afterwards planted out when these cease, and carefully cultivated to obtain blossoms in February or March.

The ZINNIA, has only of late years been introduced, but by a mistake it has generally been sown too late in the year to produce good flowers, whereas if the seed is put into the ground about June, fine handsome flowers will be the result, in the cold weather.

The CENTAURY, Centaurea, is a very pretty class of annuals which grows, and blossoms freely in this country.

The *Woolly Centaury*, C. lanata, is mentioned by Roxburgh as indigenous to the country, but the flowers are very small, of a purple color, blossoming in December.

The *Blue bottle* O. cyanus, *Azeez*, flowers in December and January, of pink and blue.

The *Sweet Sultan*, C. moschata, *Shah pusund* is known by its fragrant and delicate lilac blossoms in January and February.

The BALSAM, Impatiens, *Gulmu'hudee, doopatee* is not cultivated, or encouraged as it should be in India, where some of the varieties are indigenous. A very rich soil should be used.

Dr. R. Wight observes, that Balsams of the colder Hymalayas, like those of Europe, split from the base, rolling the segment towards the apex, whilst those of the hotter regions do the reverse.

All annuals require the same, or nearly the same treatment, of which the following may be considered a fair sketch.

Propagation.--These plants are all raised from seed put in the earth generally on the close of the rains, although some plants, such as nasturtium, sweet pea, scabious, wall-flower, and stock, are better to be sown in pots about June or July, and then put out into the border as soon as the rains cease. The seed must be sown in patches, rings, or small beds according to taste, the ground being previously stirred, and made quite fine, the earth sifted over them to a depth proportioned to the size of the seed, and then gently pressed down, so as closely to embrace every part of the seed. When the plants are an inch high they must be thinned out to a distance of two, three, five, seven, or more inches apart, according to their kind, whether spreading, or upright, having reference also to their size; the plants thinned out, if carefully taken up, may generally be transplanted to fill up any parts of the border where the seed may have failed.

Culture. Weeding and occasionally stirring the soil, and sticking such as require support, is all the cultivation necessary for annuals. If it be desired to save seed, some of the earliest and most perfect blossoms should be preserved for this purpose, so as to secure the best possible seed for the ensuing year, not leaving it to chance to gather seed from such plants as may remain after the flowers have been taken, as is generally the case with native gardeners, if left to themselves.

FLOWERS THAT GROW UNDER THE SHADE OF TREES.

It is of some value to know what these are, but at the same time it must be observed that no plant will grow under trees of the fir tribe, and it would be a great risk to place any under the *Deodar*--with all others also it must not be expected that any trees having their foliage so low as to affect the circulation of air under their branches, can do otherwise than destroy the plants placed beneath them.

Those which may be so planted are;--Wood Anemone.--Common Arum.--Deadly Nightshade--Indian ditto.--Chinese Clematis--Upright ditto--Woody Strawberry--Woody Geranium.--Green Hellebore.--Hairy St. John's Wort.-- Dog's Violet.--Imperial Fritillaria--The common Oxalis, and some other bulbs.--Common Hound's Tongue.--Common Antirrhinum.--Common Balsam.-To these may be added many of the orchidaceous plants.

ROSES.

THE ROSE, ROSA, *Gul* or *gulab*: as the most universally admired, stands first amongst shrubs. The London catalogues of this beautiful plant contain upwards of two thousand names: Mr. Loudon, in his "*Encyclopaedia of Plants*" enumerates five hundred and twenty-two, of which he describes three species, viz. Macrophylla, Brunonii, and Moschata Nepalensis, as natives of Nepal; two, viz. Involucrata, and Microphylla, as indigenous to India, and Berberifolia, and Moschata arborea, as of Persian origin, whilst twelve appear to have come from China. Dr. Roxburgh describes the following eleven species as inhabitants of these regions:--

Rosa	involucrata,
--	Chinensis,
--	semperflorens,
--	recurva,
--	microphylla,
--	inermis,
Rosa	centiflora,
--	glandulifera,
--	pubescens,
--	diffusa,
--	triphylla,

most of which, however, he represents to have been of Chinese origin.

The varieties cultivated generally in gardens are, however, all that will be here described.

These are--

1. The *Madras rose,* or *Rose Edward*, a variety of R centifolia, *Gul ssudburul*, is the most common, and has multiplied so fast within a few years, that no garden is without it, it blossoms all the year round, producing large bunches of buds at the extremities of its

shoots of the year, but, if handsome, well-shaped flowers are desired, these must be thinned out on their first appearance, to one or two, or at the most three on each stalk. It is a pretty flower, but has little fragrance. This and the other double sorts require a rich loam rather inclining to clay, and they must be kept moist.[138]

2. The *Bussorah Rose*, R gallica, *Gulsooree*, red, and white, the latter seldom met with, is one of a species containing an immense number of varieties. The fragrance of this rose is its greatest recommendation, for if not kept down, and constantly looked to, it soon gets straggling, and unsightly, like the preceding species too, the buds issue from the ends of the branches in great clusters, which must be thinned, if well formed fragrant blossoms are desired. The same soil is required as for the preceding, with alternating periods of rest by opening the roots, and of excitement by stimulating manure.

3. The *Persian rose*, apparently R collina, *Gul eeran* bears a very full-petaled blossom, assuming a darker shade as these approach nearer to the centre, but, it is difficult to obtain a perfect flower, the calyx being so apt to burst with excess of fulness, that if perfect flowers are required a thread should be tied gently round the bud, it has no fragrance. A more sandy soil will suit this kind, with less moisture.

4. The *Sweet briar* R rubiginosa, *Gul nusreen usturoon*, grows to a large size, and blossoms freely in India, but is apt to become straggling, although, if carefully clipped, it may be raised as a hedge the same as in England, it is so universally a favorite as to need no description.

5. The *China blush rose*, R Indica (R Chinensis of Roxburgh), *Kut'h gulab*, forms a pretty hedge, if carefully clipped, but is chiefly usefully as a stock for grafting on. It has no odour.

6 The *China ever-blowing rose*, R damascena of Roxburgh, *Adnee gula, gulsurkh*, bearing handsome dark crimson blossoms during the whole of the year, it is branching and bushy, but rather delicate, and wants odour.

7 The *Moss Rose*, R muscosa, having no native name is found to exist, but has only been known to have once blossomed in India;

good plants may be obtained from Hobart Town without much trouble.

8 The *Indian dog-rose*, R arvensis, R involucrata of Roxburgh, *Gul bé furman*, is found to glow wild in some parts of Nepal and Bengal, as well as in the province of Buhar, flowering in February, the blossoms large, white, and very fragrant, its cultivation extending is improving the blossoms, particularly in causing the petals to be multiplied.

9. The *Bramble-flowered rose* R multiflora, *Gul rana*, naturally a trailer, may be trained to great advantage, when it will give beautiful bunches of small many petaled flowers in February and March, of delightful fragrance.

10. The *Due de Berri rose*, a variety of R damascena, but having the petals more rounded and more regular, it is a low rather drooping shrub with delicately small branches.

Propagation.--All the species may be multiplied by seed, by layers, by cuttings, by suckers, or from grafts, almost indiscriminately. Layering is the easiest, and most certain mode of propagating this most beautiful shrub.

The roots that branch, out and throw up distinct shoots may be divided, or cut off from the main root, and even an eye thus taken off may be made to produce a good plant.

Suckers, when they have pushed through the soil, may be taken up by digging down, and gently detaching them from the roots.

Grafting or budding is used for the more delicate kinds, especially the sweet briar, and, by the curious, to produce two or more varieties on one stem, the best stocks being obtained from the China, or the Dog Rose.

Soil &c.--Any good loamy garden soil without much sand, suits the rose, but to produce it in perfection the ground can hardly be too rich.

Culture.--Immediately at the close of the rains, the branches of most kinds of roses, especially the double ones, should be cut down to not more than six inches in length, removing at the same time, all old and decayed wood, as well as all stools that have branched out from the main one, and which will form new plants; the knife being at the same time freely exercised in the removal of sickly and crowded fibres from the roots; these should likewise be laid open, cleaned and pinned, and allowed to remain exposed until blossom buds begin to appear at the end of the first shoots; the hole must then be filled with good strong stable manure, and slightly earthed over. About a month after, a basket of stable dung, with the litter, should be heaped up round the stems, and broken brick or turf placed over it to relieve the unsightly appearance.

While flowering, too, it will be well to water with liquid manure at least once a week. If it be desired to continue the trees in blossom, each shoot should be removed as soon as it has ceased flowering. To secure full large blossoms, all the buds from a shoot should be cut off, when quite young, except one.

The *Sweet briar rose* strikes its root low, and prefers shade, the best soil being a deep rich loam with very little sand, rather strong than otherwise; it will be well to place a heap of manure round the stem, above ground, covering over with turf, but it is not requisite to open the roots, or give them so much manure as for other varieties. The sweet briar must not be much pruned, overgrowth being checked rather by pinching the young shoots, or it will not blossom, and it is rather slower in throwing out shoots than other roses. In this country the best mode of multiplying this shrub is by grafting on a China rose stock, as layers do not strike freely, and cuttings cannot be made to root at all.

The *Bramble-flowered rose* is a climber, and though not needing so strong a soil as other kinds, requires it to be rich, and frequently renewed, by taking away the soil from about the roots and supplying its place with a good compost of loam, leaf mould, and well rotted dung, pruning the root. The plants require shelter from the cold wind from the North, or West, this, however, if carefully trained, they will form for themselves, but until they do so, it is impossible to make them blossom freely, the higher branches should be allowed to droop, and if growing luxuriantly, with the shoots not shortened, they will the following season, produce bunches of flowers at the end of every one, and have a very beautiful effect, no pruning should be given, except what is just enough to keep the plants within bounds, as they invariably suffer from the use of the knife. This rose is easily propagated by cuttings or layers, both of which root readily.

The *China rose* thrives almost anywhere, but is best in a soil of loam and peat, a moderate supply of water being given daily during the hot weather. They will require frequent thinning out of the branches, and are propagated by cuttings, which strike freely.[139]

As before mentioned, Rose trees look well in a parterre by themselves, but a few may be dispersed along the borders of the garden.

Insects, &c. The green, and the black plant louse are great enemies to the rose tree, and, whenever they appear, it is advisable to cut out at once the shoot attacked, the green caterpillar too, often makes skeletons of the leaves in a short time, the ladybird, as it is commonly called, is an useful insect, and worthy of encouragement, as it is a destroyer of the plant louse.

CREEPERS AND CLIMBERS

The CLIMBING, and TWINING SHRUBS offer a numerous family, highly deserving of cultivation, the following being a few of the most desirable.

The HONEY-SUCKLE, Caprifolium, having no native name, is too well known, and too closely connected with the home associations of all to need particularizing. It is remarkable that they always twine from east to west, and rather die than submit to a change.

The TRUMPET FLOWER, Bignonia, are an eminently handsome family, chiefly considered stove plants in Europe, but here growing freely in the open ground, and flowering in loose spikes.

The MOUNTAIN EBONY, Bauhinia, the distinguishing mark of the class being its two lobed leaves, most of them are indigenous, and in their native woods attain an immense size, far beyond what botanists in Europe appear to give them credit for.

The VIRGIN'S BOWER, Clematis, finds some indigenous representatives in this country, although unnamed in the native language; the odour however is rather too powerful, and of some kinds even offensive, except immediately after a shower of rain. They are all climbers, requiring the same treatment as the honey suckle.

The PASSION FLOWER, Passiflora, is a very large family of twining shrubs, many of them really beautiful, and generally of easy cultivation, this country being of the same temperature with their indigenous localities.

The RACEMOSE ASPARAGUS, A. racemosus, *Sadabooree, sutmoolee*, is a native of India, and by nature a trailing plant, but better cultivated as a climber on a trellis, in which way its delicate setaceous foliage makes it at all times ornamental, and at the close of the rains it sends forth abundant bunches of long erect spires of greenish white color, and of delicious fragrance, shedding perfume all around to a great distance.

KALENDAR WORK TO BE PERFORMED.

JANUARY.

Thin out seeding annuals wherever they appear too thick. Water freely, especially such plants as are in bloom, and keep all clean from weeds. Cut off the footstalks of flowers, except such as are reserved for seed, as soon as the petals fade. Collect the seeds of early annuals as they ripen.

FEBRUARY.

Continue as directed in last month. Prepare stocks for roses to be grafted on, R. bengalensis, and R. canina are the best. Great care must be paid to thinning out the buds of roses to insure perfect blossoms, as well as to rubbing off the succulent upright shoots and suckers that are apt to spring up at this period. Collect seeds as they ripen, to be dried, or hardened in the shade.

Collect seeds as they ripen, drying them carefully, for a few days in the pods, and subsequently when freed from them in the shade, to put them in the sun being highly injurious. Give a plentiful supply of water in saucers to Narcissus, or other bulbs when flowering.

MARCH.

Cut down the flower stalks of Narcissus that have ceased flowering, and lessen the supply of water. Take up the tubers of Dahlias, and dry gradually in an open place in the shade, but do not remove the offsets for some days. Pot any of the species of Geranium that have been put out after the rains, provided they are not in bloom. Give water freely to the roots of all flowers that are in blossom. Mignionette that is in blossom should have the seed pods clipped off with a pair of scissors every day to continue it. Convolvulus in flower should be shaded early in the morning, or it will quickly fade. The Evening Primrose should be freely watered to increase the number of blossoms. Look to the Carnations that are coming into bloom, give support to the flower stem, cutting off all side shoots and buds, except the one intended to give a handsome flower.

APRIL.

Careful watering, avoiding any wetting of the leaves is necessary at this period, and the saucers of all bulbs not yet flowered should be kept constantly full, to promote blossoming--the saucers should however be kept clean, and washed out every third day at least. Frequent weeding must be attended to, with occasional watering all grass plots, or paths. Wherever any part of the garden becomes empty by the clearing off of annuals, it should be well dug to a depth of at least eighteen inches, and after laying exposed in clods for a week or two, manured with tank or road mud; leaf mould, or other good well rotted manure.

MAY.

This is the time to make layers of Honeysuckle, Bauhinia, and other climbing and twining shrubs.

Mignionette must be very carefully treated, kept moist, and every seed- pod clipped off as soon as the flower fades, or it will not be preserved. Continue to dig, and manure the borders, not leaving the manure exposed, or it will lose power. Make pipings and layers of Carnations.

JUNE.

Thin out the multitudinous buds of the Madras rose, also examine the buds of the Persian rose, to prevent the bursting of the calyx by tying with thread, or with a piece of parchment, or cardboard as directed for Carnations.

Watch Carnations to prevent the bursting of the calyx, and to remove superfluous buds. Re pot Geraniums that are in sheds, or verandahs, so soon as they have done flowering, also take up, and pot any that may yet remain in the borders. Prune off also all superfluous, or straggling branches. Continue digging over and manuring the flowering borders. Sow Zinnias, also make cuttings of perennials and biennials that are propagated by that means, and put in seeds of biennials under shelter, as well as a few of the early annuals, particularly Stock and Sweet-pea.

JULY.

Make cuttings and layers of hardy shrubs, and of the Fragrant Olive; put in cuttings of the Willow, and some other trees. Plant out Pines, and Casuarina, Cypress, Large-leaved fig, and the Laurel tribe. Transplant young shrubs of a hardy nature.

Divide the roots, and plant out suckers, or offsets of perennial border plants. Make cuttings and sow seeds of biennials, as required; also a few annuals to be hereafter transplanted. Sow also Geraniums. Continue making pipings of Carnation, plant out, or transplant hardy perennials into the borders.

AUGUST.

This may be considered the best time for sowing the seeds of hardy shrubs. Plant out Aralia, Canella, Magnolia, and other ornamental trees. Transplant delicate and exotic shrubs. Remove, and plant out suckers, and layers of hardy shrubs. Prune all shrubs freely.

Divide, and plant out suckers, and offsets of hardy perennials, that have formed during the rains. Plant out tender perennial plants, in the borders, also biennials. Prune, and thin out perennial plants in the borders. Put out in the borders such annuals as were sown in June, protecting them from the heat of the sun in the afternoon. Sow a few early annuals. Plant out Dahlia tubers where they are intended to blossom, keeping them as much as possible in classes of colors. Make pipings of Carnations.

SEPTEMBER.

Prick out the cuttings of hardy shrubs that have been made before, or during the rains, in beds for growing. Prune all flowering shrubs, having due regard to the character of each, as bearing flowers on the end of the shoots, or from the side exits, give the annual dressing of manure to the entire shrubbery, with new upper soil.

Remove the top soil from the borders, and renew with addition of a moderate quantity of manure. Put out Geraniums into the borders, and set rooted cuttings singly in pots. Plant out biennials in the borders, also such annuals as have been sown in pots. Re-pot and give fresh earth to plants in the shed.

OCTOBER.

Open out the roots of a few Bussorah roses for early flowering, pruning down all the branches to a height of six inches, removing all decayed, and superannuated wood, dividing the roots, and pruning them freely. The Madras roses should be treated in the same manner, not all at the same time, but at intervals of a week between each cutting down, so as to secure a succession for blossoming. Plant out rooted cuttings in beds, to increase in size.

Sow annuals freely, and thin out those put in last month, so as to leave sufficient space for growing, at the same time transplanting the most healthy to other parts of the border.

NOVEMBER.

Continue opening the roots of Bussorah roses, as well as the Rose Edward, and Madras roses, for succession to those on which this operation was performed last month. Prune, and trim the Sweetbriar, and Many-flowered rose.

Flower-Garden--Divide, and plant bulbs of all kinds, both, for border, and pot flowering. Continue to sow annuals.

DECEMBER

Continue opening the roots, and cutting down the branches of Bussorah, and other roses for late flowering. Prune, and thin out also the China and Persian roses, as well as the Many-flowered rose, if not done last month. Train carefully all climbing and twining shrubs.

Weed beds of annuals, and thin out, where necessary. Sow Nepolitan, and other fine descriptions of Larkspur, as well as all other annuals for a late show. Dahlias are now blooming in perfection, and should be closely watched that every side-bud, or more than one on each stalk may be cut off close, with a pair of scissors to secure full, distinctly colored, and handsome flowers.

[For further instructions respecting the culture of flowers in India I must refer my readers to the late Mr. Speede's works, where they will find a great deal of useful information not only respecting the flower- garden, but the kitchen-garden and the orchard.]

MISCELLANEOUS ITEMS.

THE TREE-MIGNONETTE.--This plant does not appear to be a distinct variety, for the common mignonette, properly trained becomes shrubby. It may be propagated by either seed or cuttings. When it has put forth four leaves or is about an inch high, take it from the bed and put it by itself into a moderate sized pot. As it advances in growth, carefully pick off all the side shoots, leaving the leaf at the base of each shoot to assist the growth of the plant. When it has reached a foot in height it will show flower. But every flower must be nipped off carefully. Support the stem with a stick to make it grow straight. Even when it has attained its proper height of two feet again cut off the bloom for a few days.

It is said that Miss Mitford, the admired authoress, was the first to discover that the common mignonette could be induced to adopt tree-like habits. The experiment has been tried in India, but it has sometimes failed from its being made at the wrong season. The seed should be sown at the end of the rains.

GRAFTING.--Take care to unite exactly the inner bark of the scion with the inner bark of the stock in order to facilitate the free course of the sap. Almost any scion will take to almost any sort of tree or plant provided there be a resemblance in their barks. The Chinese are fond of making fantastic experiments in grafting and sometimes succeed in the most heterogeneous combinations, such as grafting flowers upon fruit trees. Plants growing near each other can sometimes be grafted by the roots, or on the living root of a tree cut down another tree can be grafted. The scions are those shoots which united with the stock form the graft. It is desirable that the sap of the stock should be in brisk and healthy motion at the time of grafting. The graft should be surrounded with good stiff clay with a little horse or cow manure in it and a portion of cut hay. Mix the materials with a little water and then beat them up with a stick until the compound is quite ductile. When applied it may be bandaged with a cloth. The best season for grafting in India is the rains.

MANURE.--Almost any thing that rots quickly is a good manure. It is possible to manure too highly. A plant sometimes dies from too much richness of soil as well as from too barren a one.

WATERING.--Keep up a regular moisture, but do not deluge your plants until the roots rot. Avoid giving very cold water in the heat of the day or in the sunshine. Even in England some gardeners in a hot summer use luke-warm water for delicate plants. But do not in your fear of overwatering only wet the surface. The earth all round and below the root should be equally moist, and not one part wet and the other dry. If the plant requires but little water, water it seldom, but let the water reach all parts of the root equally when you water at all.

GATHERING AND PRESERVING FLOWERS.--Always use the knife, and prefer such as are coming into flower rather than such as are fully expanded. If possible gather from crowded plants, or parts of plants, so that every gathering may operate at the same time as a judicious pruning and thinning. Flowers may be preserved when gathered, by inserting their ends in winter, in moist earth, or moss; and may be freshened, when withered, by sprinkling them with water, and putting them in a close vessel, as under a bellglass, handglass, flowerpot or in a botanic box; if this will not do, sprinkle them with warm water heated to 80° or 90°, and cover them with a glass.--*Loudon's Encyclopaedia of Gardening*.

PIPING---is a mode of propagation by cuttings and is adopted in plants having joined tubular stems, as the dianthus tribe. When the shoot has nearly done growing (soon after its blossom has fallen) its extremity is to be separated at a part of the stem where it is hard and ripe. This is done by holding the root with one hand and with the other pulling the top part above the pair of leaves so as to separate it from the root part of the stem at the socket, formed by the axillae of the leaves, leaving the stem to remain with a tubular or pipe-looking termination. The piping is inserted in finely sifted earth to the depth of the first joint or pipe and its future management regulated on the same general principles as cuttings.--*From the same*.

BUDDING.--This is performed when the leaves of plants have grown to their full size and the bud is to be seen at the base of it. The relative nature of the bud and the stock is the same as in grafting. Make a slit in the bark of the stock, to reach from half an inch to an inch and a half down the stock, according to the size of the plant; then make another short slit across, that you may easily raise the bark from the wood, then take a very thin slice of the bark from the tree or plant to be budded, a little below a leaf, and bring the knife out a little above it, so that you remove the leaf and the bud at its base, with the little slice you have taken. You will perhaps have removed a small bit of the wood with the bark, which you must take carefully out with the sharp point of your knife and your thumb; then tuck the bark and bud under the bark of the stock which you carefully bind over, letting the bud come at the part where the slits cross each other. No part of the stock should be allowed to grow after it is budded, except a little shoot or so, above the bud, just to draw the sap past the bud.--*Gleenny's Hand Book of Gardening*.

ON PYRAMIDS OF ROSES.--The standard Roses give a fine effect to a bed of Roses by being planted in the middle, forming a pyramidal bed, or alone on grass lawns; but the *ne plus ultra* of a pyramid of Roses is that formed of from one, two, or three plants, forming a pyramid by being trained up three strong stakes, to any length from 10 to 25 feet high (as may suit situation or taste), placed about two feet apart at the bottom; three forming an angle on the ground, and meeting close together at the top; the plant, or plants to be planted inside the stakes. In two or three years, they will form a pyramid of Roses which baffles all description. When gardens are small, and the owners are desirous of having *multum in parvo*, three or four may be planted to form one pyramid; and this is not the only object of planting more sorts than one together, but the beauty is also much increased by the mingled hues of the varieties planted. For instance, plant together a white Boursault, a purple Noisette, a Stadtholder, Sinensis (fine pink), and a Moschata scandens and such a variety may be obtained, that twenty pyramids may have each, three or four kinds, and no two sorts alike on the whole twenty pyramids. A temple of Roses, planted in the same way, has a beautiful appearance in a flower garden--that is, eight, ten, or twelve stout peeled Larch poles, well painted, set in the ground, with a light iron rafter from each, meeting at the top and forming a dome. An old cable, or other old rope, twisted round the pillar and iron, gives an additional beauty to the whole. Then plant against the pillars with two or three varieties, each of which will soon run up the pillars, and form a pretty mass of Roses, which amply repays the trouble and expense, by the elegance it gives to the garden--*Floricultural Cabinet.*

How TO MAKE ROSE WATER, &c--Take an earthen pot or jar well glazed inside, wide in the month, narrow at the bottom, about 15 inches high, and place over the mouth a strainer of clean coarse muslin, to contain a considerable quantity of rose leaves, of some highly fragrant kind. Cover them with a second strainer of the same material, and close the mouth of the jar with an iron lid, or tin cover, hermetically sealed. On this lid place hot embers, either of coal or charcoal, that the heat may reach the rose-leaves without scorching or burning them.

The aromatic oil will fall drop by drop to the bottom with the water contained in the petals. When time has been allowed for extracting the whole, the embers must be removed, and the vase placed in a cool spot.

Rose-water obtained in this mode is not so durable as that obtained in the regular way by a still but it serves all ordinary purposes. Small alembics of copper with a glass capital, may be used in three different ways.

In the first process, the still or alembic must be mounted on a small brick furnace, and furnished with a worm long enough to pass through a pan of cold water. The petals of the rose being carefully picked so as to leave no extraneous parts, should be thrown into the boiler of the still with a little water.

The great point is to keep up a moderate fire in the furnace, such as will cause the vapour to rise without imparting a burnt smell to the rose water.

The operation is ended when the rose water, which falls drop by drop in the tube, ceases to be fragrant. That which is first condensed has very little scent, that which is next obtained is the best, and the third and last portion is generally a little burnt in smell, and bitter in taste. In a very small still, having no worm, the condensation must be produced by linen, wetted in cold water, applied round the capital. A third method consists in plunging the boiler of the still into a larger vessel of boiling water placed over a fire, when the rose-water never acquires the burnt flavour to which we have alluded. By another process, the still is placed in a boiler filled with sand instead of water, and heated to the necessary temperature.

But this requires alteration, or it is apt to communicate a baked flavour.

SYRUP OF ROSES--May be obtained from Belgian or monthly roses, picked over, one by one, and the base of the petal removed. In a China Jar prepared with a layer of powdered sugar, place a layer of rose-leaves about half an inch thick; then of sugar, then of leaves, till the vessel is full.

On the top, place a fresh wooden cover, pressed down with a weight. By degrees, the rose-leaves produce a highly-coloured, highly-scented syrup; and the leaves form a colouring-matter for liqueurs.

PASTILLES DU SERAIL.--Sold in France as Turkish, in rosaries and other ornaments, are made of the petals of the Belgian or Puteem Rose, ground to powder and formed into a paste by means of liquid gum.

Ivory-black is mixed with the gum to produce a black colour; and cinnabar or vermilion, to render the paste either brown or red.

It may be modelled by hand or in a mould, and when dried in the sun, or a moderate oven, attains sufficient hardness to be mounted in gold or silver.--*Mrs. Gore's Rose Fancier's Manual*.

OF FORMING AND PRESERVING HERBARIUMS.--The most exact descriptions, accompanied with the most perfect figures, leave still something to be desired by him who wishes to know completely a natural being. This nothing can supply but the autopsy or view of the object itself. Hence the advantage of being able to see plants at pleasure, by forming dried collections of them, in what are called herbariums.

A good practical botanist, Sir J.E. Smith observes, must be educated among the wild scenes of nature, while a finished theoretical one requires the additional assistance of gardens and books, to which must be superadded the frequent use of a good herbarium. When plants are well dried, the original forms and positions of even their minutest parts, though not their colours, may at any time be restored by immersion in hot water. By this means the productions of the most distant and various countries, such as no garden could possibly supply, are brought together at once under our eyes, at any season of the year. If these be assisted with drawings and descriptions, nothing less than an actual survey of the whole vegetable world in a state of nature, could excel such a store of information.

With regard to the mode or state in which plants are preserved, desiccation, accompanied by pressing, is the most generally used. Some persons, Sir J.E. Smith observes, recommend the preservation of specimens in weak spirits of wine, and this mode is by far the most eligible for such as are very juicy: but it totally destroys their colours, and often renders their parts less fit for examination than by the process of drying. It is, besides, incommodious for frequent study, and a very expensive and bulky way of making an herbarium.

The greater part of plants dry with facility between the leaves of books, or other paper, the smoother the better. If there be plenty of paper, they often dry best without shifting; but if the specimens are crowded, they must be taken out frequently, and the paper dried before they are replaced. The great point to be attended to is, that the process should meet with no check. Several vegetables are so tenacious of their vital principle, that they will grow between papers; the consequence of which is, a destruction of their proper habit and colors. It is necessary to destroy the life of such, either by immersion in boiling water or by the application of a hot iron, such as is used for linen, after which they are easily dried. The practice of applying such an iron, as some persons do, with great labor and perseverance, till the plants are quite dry, and all their parts incorporated into a smooth flat mass is not approved of. This renders them unfit for subsequent examination, and destroys their natural habit, the most important thing to be preserved. Even in spreading plants between papers, we should refrain from that practice and artificial disposition of their branches, leaves, and other parts, which takes away from their natural aspect, except for the purpose of displaying the internal parts of some one or two of their flowers, for ready observation. The most approved method of pressing is by a box or frame, with a bottom of cloth or leather, like a square sieve. In this, coarse sand or small shot may be placed; in any quantity very little pressing is required in drying specimens; what is found necessary should be applied equally to every part of the bundle under the operation.

Hot-pressing, by means of steel net-work heated, and placed in alternate layers with the papers, in the manner of hot pressing paper, and the whole covered with the equalizing press, above described, would probably be an improvement, but we have not heard of its being tried. At all events, pressing by screw presses, or weighty non-elastic bodies, must be avoided, as tending to bruise the stalks and other protuberant parts of plants.

"After all we can do," Sir J.E. Smith observes, "plants dry very variously. The blue colours of their flowers generally fade, nor are reds always permanent. Yellows are much more so, but very few white flowers retain their natural aspect. The snowdrop and parnassia, if well dried, continue white. Some greens are much more permanent than others; for there are some natural families whose leaves, as well as flowers, turn almost black by drying, as melampyrum, bartsia, and their allies, several willows, and most of the orchideae. The heaths and firs in general cast off their leaves between papers, which appears to be an effort of the living principle, for it is prevented by immersion of the fresh specimen in boiling water."

The specimens being dried, are sometimes kept loose between leaves of paper; at other times wholly gummed or glued to paper, but most generally attached by one or more transverse slips of paper, glued on one end and pinned at the other, so that such specimens can readily be taken out, examined, and replaced. On account of the aptitude of the leaves and other parts of dried plants to drop off, many glue them entirely, and such seems to be the method adopted by Linnaeus, and recommended by Sir J.E. Smith. "Dried specimens," the professor observes, "are best preserved by being fastened, with weak carpenter's glue, to paper, so that they may be turned over without damage. Thick and heavy stalks require the additional support of a few transverse strips of paper, to bind them more firmly down. A half sheet, of a convenient folio size, should be allotted to each species, and all the species of a genus may be placed in one or more whole sheets or folios. On the latter outside should be written the name of the genus, while the name of every species, with its place of growth, time of gathering, the finder's name, or any other concise piece of information, may be inscribed on its appropriate paper. This is the plan of the Linnaean herbarium."--*Loudon*.

THE END.

FOOTNOTES.

[001] Some of the finest *Florists flowers* have been reared by the mechanics of Norwich and Manchester and by the Spitalfield's weavers. The pitmen in the counties of Durham and Northumberland reside in long rows of small houses, to each of which is attached a little garden, which they cultivate with such care and success, that they frequently bear away the prize at Floral Exhibitions.

[002] Of Rail-Road travelling the reality is quite different from the idea that descriptions of it had left upon my mind. Unpoetical as this sort of transit may seem to some minds, I confess I find it excite and satisfy the imagination. The wondrous speed--the quick change of scene-- the perfect comfort--the life-like character of the power in motion, the invisible, and mysterious, and mighty steam horse, urged, and guided, and checked by the hand of Science--the cautionary, long, shrill whistle--the beautiful grey vapor, the breath of the unseen animal, floating over the fields by which we pass, sometimes hanging stationary for a moment in the air, and then melting away like a vision--furnish sufficiently congenial amusement for a period-minded observer.

[003] "That which peculiarly distinguishes the gardens of England," says Repton, "is the beauty of English verdure: *the grass of the mown lawn*, uniting with, the grass of the adjoining pastures, and presenting *that permanent verdure* which is the natural consequence of our soft and humid clime, but unknown to the cold region of the

North or the parching temperature of the South. This it is impossible to enjoy in Portugal where it would be as practicable to cover the general surface with the snow of Lapland as with the verdure of England." It is much the same in France. "There is everywhere in France," says Loudon, "a want *of close green turf*, of ever-green bushes and of good adhesive gravel." Some French admirers of English gardens do their best to imitate our lawns, and it is said that they sometimes partially succeed with English grass seed, rich manure, and constant irrigation. In Bengal there is a very beautiful species of grass called Doob grass, (*Panicum Dactylon*,) but it only flourishes on wide and exposed plains with few trees on them, and on the sides of public roads, Shakespeare makes Falstaff say that "the camomile the more it is trodden on the faster it grows" and, this is the case with the Doob grass. The attempt to produce a permanent Doob grass lawn is quite idle unless the ground is extensive and open, and much trodden by men or sheep. A friend of mine tells me that he covered a large lawn of the coarse Ooloo grass (*Saccharum cylindricum*) with mats, which soon killed it, and on removing the mats, the finest Doob grass sprang up in its place. But the Ooloo grass soon again over-grew the Doob.

[004] I allude here chiefly to the ryots of wealthy Zemindars and to other poor Hindu people in the service of their own countrymen. All the subjects of the British Crown, even in India, are *politically free*, but individually the poorer Hindus, (especially those who reside at a distance from large towns,) are unconscious of their rights, and even the wealthier classes have rarely indeed that proud and noble feeling of personal independence which characterizes people of all classes and conditions in England. The feeling with which even a Hindu of wealth and rank approaches a man in power is very different indeed from that of the poorest Englishman under similar circumstances. But national education will soon communicate to the natives of India a larger measure of true self-respect. It will not be long, I hope, before the Hindus will understand our favorite maxim of English law, that "Every man's house is his castle,"--a maxim so finely amplified by Lord Chatham: "*The poorest man may in his cottage bid defiance to all the forces of the Crown. It may be frail--its roof may shake--the wind may blow through it--the storm may enter--but*

the king of England cannot enter!--all his force dares not cross the threshold of the ruined tenement."

[005]*Literary Recreations.*

[006] I have in some moods preferred the paintings of our own Gainsborough even to those of Claude--and for this single reason, that the former gives a peculiar and more touching interest to his landscapes by the introduction of sweet groups of children. These lovely little figures are moreover so thoroughly English, and have such an out-of- doors air, and seem so much a part of external nature, that an Englishman who is a lover of rural scenery and a patriot, can hardly fail to be enchanted with the style of his celebrated countryman.--*Literary Recreations.*

[007] Had Evelyn only composed the great work of his 'Sylva, or a Discourse of Forest Trees,' &c. his name would have excited the gratitude of posterity. The voice of the patriot exults in his dedication to Charles II, prefixed to one of the later editions:--'I need not acquaint your Majesty, how many millions of timber-trees, besides infinite others, have been propagated and planted throughout your vast dominions, at the instigation and by the sole direction of this work, because your Majesty has been pleased to own it publicly for my encouragement.' And surely while Britain retains her awful situation among the nations of Europe, the 'Sylva' of Evelyn will endure with her triumphant oaks. It was a retired philosopher who aroused the genius of the nation, and who casting a prophetic eye towards the age in which we live, has contributed to secure our sovereignty of the seas. The present navy of Great Britain has been constructed with the oaks which the genius of Evelyn planted.--*D'Israeli's Curiosities of Literature.*

[008]*Crisped knots* are figures curled or twisted, or having waving lines intersecting each other. They are sometimes planted in box. Children, even in these days, indulge their fancy in sowing mustard and cress, &c. in 'curious knots,' or in favorite names and sentences. I have done it myself, "I know not how oft,"--and alas, how long ago! But I still remember with what anxiety I watered and watched the ground, and with what rapture I at last saw the surface gradually rising and breaking on the light green heads of the delicate little

new-born plants, all exactly in their proper lines or stations, like a well- drilled Lilliputian battalion.

Shakespeare makes mention of garden *knots* in his *Richard the Second*, where he compares an ill governed state to a neglected garden.

> Why should we, in the compass of a pale, Keep law, and form, and due proportion, Showing, as in a model, our firm estate? When our sea-walled garden, the whole land, Is full of weeds; her finest flowers choked up, Her fruit-trees all unpruned, her hedges ruined, Her *knots* disordered, and her wholesome herbs Swarming with caterpillars.

There is an allusion to garden *knots* in *Holinshed's Chronicle*. In 1512 the Earl of Northumberland "had but one gardener who attended hourly in the garden for setting of erbis and *chipping of knottis* and sweeping the said garden clean."

[009] Ovid, in his story of Pyramus and Thisbe, tells us that the black Mulberry was originally white. The two lovers killed themselves under a white Mulberry tree and the blood penetrating to the roots of the tree mixed with the sap and gave its color to the fruit.

[010] *Revived Adonis*,--for, according to tradition he died every year and revived again. *Alcinous, host of old Laertes' son*,--that is, of Ulysses, whom he entertained on his return from Troy. *Or that, not mystic*--not fabulous as the rest, but a real garden which Solomon made for his wife, the daughter of Pharoah, king of Egypt--WARBURTON

"Divested of harmonious Greek and bewitching poetry," observes Horace Walpole, "the garden of Alcinous was a small orchard and vineyard with some beds of herbs and two fountains that watered them, inclosed within a quickset hedge." Lord Kames, says, still more boldly, that it was nothing but a kitchen garden. Certainly, gardening amongst the ancient Greeks, was a very simple business. It is only within the present century that it has been any where elevated into a fine art.

[011] "We are unwilling to diminish or lose the credit of Paradise, or only pass it over with [the Hebrew word for] *Eden*, though the Greek be of a later name. In this excepted, we know not whether the

ancient gardens do equal those of late times, or those at present in Europe. Of the gardens of Hesperides, we know nothing singular, but some golden apples. Of Alcinous his garden, we read nothing beyond figs, apples, olives; if we allow it to be any more than a fiction of Homer, unhappily placed in Corfu, where the sterility of the soil makes men believe there was no such thing at all. The gardens of Adonis were so empty that they afforded proverbial expression, and the principal part thereof was empty spaces, with herbs and flowers in pots. I think we little understand the pensile gardens of Semiramis, which made one of the wonders of it [Babylon], wherein probably the structure exceeded the plants contained in them. The excellency thereof was probably in the trees, and if the descension of the roots be equal to the height of trees, it was not [absurd] of Strebæus to think the pillars were hollow that the roots might shoot into them."--*Sir Thomas Browne.--Bohn's Edition of Sir Thomas Browne's Works, vol. 2, page* 498.

[012] The house and garden before Pope died were large enough for their owner. He was more than satisfied with them. "As Pope advanced in years," says Roscoe, "his love of gardening, and his attention to the various occupations to which it leads, seem to have increased also. This predilection was not confined to the ornamental part of this delightful pursuit, in which he has given undoubted proofs of his proficiency, but extended to the useful as well as the agreeable, as appears from several passages in his poems; but he has entered more particularly into this subject in a letter to Swift (March 25, 1736); "I wish you had any motive to see this kingdom. I could keep you: for I am rich, that is, have more than I want, I can afford room to yourself and two servants. I have indeed room enough; nothing but myself at home. The kind and hearty housewife is dead! The agreeable and instructive neighbour is gone! Yet my house is enlarged, and the gardens extend and flourish, as knowing nothing of the guests they have lost. I have more fruit trees and kitchen garden than you have any thought of; and, I have good melons and apples of my own growth. I am as much a better gardener, as I am a worse poet, than when you saw me; but gardening is near akin to philosophy, for Tully says, *Agricultura proxima sapientiae*. For God's sake, why should not you, (that are a step higher than a philosopher, a divine, yet have too much grace and wit than

to be a bishop) even give all you have to the poor of Ireland (for whom you have already done every thing else,) so quit the place, and live and die with me? And let *tales anima concordes* be our motto and our epitaph."

[013] The leaves of the willow, though green above, are hoar below. Shakespeare's knowledge of the fact is alluded to by Hazlitt as one of the numberless evidences of the poet's minute observation of external nature.

[014] See Mr. Loudon's most interesting and valuable work entitled *Arboretum et Fruticetum Britanicum*.

[015] All the rules of gardening are reducible to three heads: the contrasts, the management of surprises and the concealment of the bounds. "Pray, what is it you mean by the contrasts?" "The disposition of the lights and shades."--"'Tis the colouring then?"--"Just that."--"Should not variety be one of the rules?"--"Certainly, one of the chief; but that is included mostly in the contrasts." I have expressed them all in two verses[140] (after my manner, in very little compass), which are in imitation of Horace's--*Omne tulit punctum. Pope.--Spence's Anecdotes.*

[016] In laying out a garden, the chief thing to be considered is the genius of the place. Thus at Tiskins, for example, Lord Bathurst should have raised two or three mounts, because his situation is *all* plain, and nothing can please without variety. *Pope--Spence's Anecdotes.*

[017] The seat and gardens of the Lord Viscount Cobham, in Buckinghamshire. Pope concludes the first Epistle of his Moral Essays with a compliment to the patriotism of this nobleman.

> And you, brave Cobham! to the latest breath Shall feel your ruling passion strong in death: Such in those moments as in all the past "Oh, save my country, Heaven!" shall be your last.

[018] Two hundred acres and two hundred millions of francs were made over to Le Notre by Louis XIV. to complete these geometrical gardens. One author tells us that in 1816 the ordinary cost of putting a certain portion of the waterworks in play was at the rate

of 200 £. per hour, and another still later authority states that when the whole were set in motion once a year on some Royal fête, the cost of the half hour during which the main part of the exhibition lasted was not less than 3,000 £. This is surely a most senseless expenditure. It seems, indeed, almost incredible. I take the statements from *Loudon's* excellent *Encyclopaedia of Gardening*. The name of one of the original reporters is Neill; the name of the other is not given. The gardens formerly were and perhaps still are full of the vilest specimens of verdant sculpture in every variety of form. Lord Kames gives a ludicrous account of the vomiting stone statues there;-- "A lifeless statue of an animal pouring out water may be endured" he observes, "without much disgust: but here the lions and wolves are put in violent action; each has seized its prey, a deer or a lamb, in act to devour; and yet, as by hocus-pocus, the whole is converted into a different scene: the lion, forgetting his prey, pours out water plentifully; and the deer, forgetting its danger, performs the same work: a representation no less absurd than that in the opera, where Alexander the Great, after mounting the wall of a town besieged, turns his back to the enemy, and entertains his army with a song."

[019] Broome though a writer of no great genius (if any), had yet the honor to be associated with Pope in the translation of the Odyssey. He translated the 2nd, 6th, 8th, 11th, 16th, 18th, and 23rd books. Henley (Orator Henley) sneered at Pope, in the following couplet, for receiving so much assistance:

> Pope came clean off with Homer, but they say, Broome went before, and kindly swept the way.

Fenton was another of Pope's auxiliaries. He translated the 1st, 4th, 19th and 20th books (of the Odyssey). Pope himself translated the rest.

[020] Stowe

[021] The late Humphrey Repton, one of the best landscape-gardeners that England has produced, and who was for many years employed on alterations and improvements in the house and grounds at Cobham, in Kent, the seat of the Earl of Darnley, seemed to think that Stowe ought not to monopolize applause and admira-

tion, "Whether," he said, "we consider its extent, its magnificence or its comfort, there are few places that can vie with Cobham." Repton died in 1817, and his patron and friend the Earl of Darnley put up at Cobham an inscription to his memory.

The park at Cobham extends over an area of no less than 1,800 acres, diversified with thick groves and finely scattered single trees and gentle slopes and broad smooth lawns. Some of the trees are singularly beautiful and of great age and size. A chestnut tree, named the Four Sisters, is five and twenty feet in girth. The mansion, of which, the central part was built by Inigo Jones, is a very noble one. George the Fourth pronounced the music room the finest room in England. The walls are of polished white marble with pilasters of sienna marble. The picture gallery is enriched with valuable specimens of the genius of Titian and Guido and Salvator Rosa and Sir Joshua Reynolds. There is another famous estate in Kent, Knole, the seat of

> Dorset, the grace of courts, the Muse's pride.

The Earl of Dorset, though but a poetaster himself, knew how to appreciate the higher genius of others. He loved to be surrounded by the finest spirits of his time. There is a pleasant anecdote of the company at his table agreeing to see which amongst them could produce the best impromptu. Dryden was appointed arbitrator. Dorset handed a slip of paper to Dryden, and when all the attempts were collected, Dryden decided without hesitation that Dorset's was the best. It ran thus: "*I promise to pay Mr. John Dryden, on demand, the sum of £500. Dorset.*"

[022] This is generally put into the mouth of Pope, but if we are to believe Spence, who is the only authority for the anecdote, it was addressed to himself.

[023] It has been said that in laying out the grounds at Hagley, Lord Lyttelton received some valuable hints from the author of *The Seasons,* who was for some time his Lordship's guest. The poet has commemorated the beauties of Hagley Park in a description that is familiar to all lovers of English poetry. I must make room for a few of the concluding lines.

> Meantime you gain the height, from whose fair brow, The bursting prospect spreads immense around: And snatched o'er hill, and dale, and wood, and lawn, And verdant field, and darkening heath between, And villages embosomed soft in trees, And spiry towns by surging columns marked, Of household smoke, your eye excursive roams; Wide stretching from the hall, in whose kind haunt The hospitable genius lingers still, To where the broken landscape, by degrees, Ascending, roughens into rigid hills; O'er which the Cambrian mountains, like far clouds, That skirt the blue horizon, dusky rise.

It certainly does not look as if there had been any want of kindly feeling towards Shenstone on the part of Lyttelton when we find the following inscription in Hagley Park.

> To the memory of William Shenstone, Esquire, In whose verse Were all the natural graces. And in whose manners Was all the amiable simplicity Of pastoral poetry, With the sweet tenderness Of the elegiac.

There is also at Hagley a complimentary inscription on an urn to Alexander Pope; and, on an octagonal building called *Thomson's Seat*, there is an inscription to the author of *The Seasons*. Hagley is kept up with great care and is still in possession of the descendants of the founder. But a late visitor (Mr. George Dodd) expresses a doubt whether the Leasowes, even in its comparative decay, is not a finer bit of landscape, a more delightful place to lose one-self in, than even its larger and better preserved neighbour.

[024] Coleridge is reported to have said--"There is in Crabbe an absolute defect of high imagination; he gives me little pleasure. Yet no doubt he has much power of a certain kind, and it is good to cultivate, even at some pains, a catholic taste in literature." Walter Savage Landor, in his "Imaginary Conversations," makes Porson say--"Crabbe wrote with a two-penny nail and scratched rough truths and rogues' facts on mud walls." Horace Smith represents Crabbe, as "Pope in worsted stockings." That there is merit of some sort or other, and that of no ordinary kind, in Crabbe's poems, is

what no one will deny. They relieved the languor of the last days of two great men, of very different characters--Sir Walter Scott and Charles James Fox.

[025] The poet had a cottage and garden in Kew-foot-Lane at or near Richmond. In the alcove in the garden is a small table made of the wood of the walnut tree. There is a drawer to the table which in all probability often received charge of the poet's effusions hot from the brain. On a brass tablet inserted in the top of the table is this inscription--"*This table was the property of James Thomson, and always stood in this seat.*"

[026] Shene or Sheen: the old name of Richmond, signifying in Saxon *shining* or *splendour*.

[027] Highgate and Hamstead.

[028] In his last sickness

[029] On looking back at page 36 I find that I have said in the foot note that it is only within *the present century* that gardening has been elevated into *a fine art*. I did not mean within the 55 years of this 19th century, but *within a hundred years*. Even this, however, was an inadvertency. We may go a little further back. Kent and Pope lived to see Landscape-Gardening considered a fine art. Before their time there were many good practical gardeners, but the poetry of the art was not then much regarded except by a very few individuals of more than ordinary refinement.

[030] Catherine the Second grossly disgraced herself as a woman--partly driven into misconduct herself by the behaviour of her husband--but as a sovereign it cannot be denied that she exhibited a penetrating sagacity and great munificence; and perhaps the lovers of literature and science should treat her memory with a little consideration. When Diderot was in distress and advertized his library for sale, the Empress sent him an order on a banker at Paris for the amount demanded, namely fifteen thousand livres, on condition that the library was to be left as a deposit with the owner, and that he was to accept a gratuity of one thousand livres annually for taking charge of the books, until the Empress should require them. This was indeed a delicate and ingenious kindness. Lord Brougham makes D'Alembert and not Diderot the subject of this

anecdote. It is a mistake. See the Correspondence of Baron de Gumm and Diderot with the Duke of Saxe-Gotha.

Many of the Russian nobles keep up to this day the taste in gardening introduced by Catherine the Second, and have still many gardens laid out in the English style. They have often had in their employ both English and Scottish gardeners. There is an anecdote of a Scotch gardener in the Crimea in one of the public journals:--

"Our readers"--says the *Banffshire Journal*--"will recollect that when the Allies made a brief expedition to Yalto, in the south of the Crimea, they were somewhat surprised and gratified by the sight of some splendid gardens around a seat of Prince Woronzow. Little did our countrymen think that these gardens were the work of a Scotchman, and a Moray loon; yet such was the case." The history of the personage in question is a somewhat singular one: "Jamie Sinclair, the garden boy, had a natural genius, and played the violin. Lady Cumming had this boy educated by the family tutor, and sent him to London, where he was well known in 1836-7-8, for his skill in drawing and colouring. Mr. Knight, of the Exotic Nursery, for whom he used to draw orchids and new plants, sent him to the Crimea, to Prince Woronzow, where he practised for thirteen years. He had laid out these beautiful gardens which the allies the other day so much admired; had the care of 10,000 acres of vineyards belonging to the prince; was well known to the Czar, who often consulted him about improvements, and gave him a "medal of merit" and a diploma or passport, by which he was free to pass from one end of the empire to the other, and also through Austria and Prussia, I have seen these instruments. He returned to London in 1851, and was just engaged with a London publisher for a three years' job, when Menschikoff found the Turks too hot for him last April twelve-month; the Russians then made up for blows, and Mr. Sinclair was more dangerous for them in London than Lord Aberdeen. He was the only foreigner who was ever allowed to see all that was done in and out of Sebastopol, and over all the Crimea. The Czar, however, took care that Sinclair could not join the allies; but where he is and what he is about I must not tell, until the war is over--except that he is not in Russia, and that he will never play first fiddle again in Morayshire."

[031] Brown succeeded to the popularity of Kent. He was nicknamed, *Capability Brown*, because when he had to examine grounds previous to proposed alterations and improvements he talked much of their *capabilities*. One of the works which are said to do his memory most honor, is the Park of Nuneham, the seat of Lord Harcourt. The grounds extend to 1,200 acres. Horace Walpole said that they contained scenes worthy of the bold pencil of Rubens, and subjects for the tranquil sunshine of Claude de Lorraine. The following inscription is placed over the entrance to the gardens.

> Here universal Pan, Knit with the Graces and the Hours in dance, Leads on the eternal Spring.

It is said that the *gardens* at Nuneham were laid out by Mason, the poet.

[032] Mrs. Stowe visited the Jardin Mabille in the Champs Elysées, a sort of French Vauxhall, where small jets of gas were so arranged as to imitate "flowers of the softest tints and the most perfect shape."

[033] Napoleon, it is said, once conceived the plan of roofing with glass the gardens of the Tuileries, so that they might be used as a winter promenade.

[034] Addison in the 477th number of the *Spectator* in alluding to Kensington Gardens, observes; "I think there are as many kinds of gardening as poetry; our makers of parterres and flower gardens are epigrammatists and sonnetteers in the art; contrivers of bowers and grottos, treillages and cascades, are romance writers. Wise and London are our heroic poets; and if I may single out any passage of their works to commend I shall take notice of that part in the upper garden at Kensington, which was at first nothing but a gravel pit. It must have been a fine genius for gardening that could have thought of forming such an unsightly hollow unto so beautiful an area and to have hit the eye with so uncommon and agreeable a scene as that which it is now wrought into."

[035] Lord Bathurst, says London, informed Daines Barrington, that *he* (Lord Bathurst) was the first who deviated from the straight line in sheets of water by following the lines in a valley in widening a brook at Ryskins, near Colnbrook; and Lord Strafford, thinking that it was done from poverty or economy asked him to own fairly how little more it would have cost him to have made it straight. In these days no possessor of a park or garden has the water on his grounds either straight or square if he can make it resemble the Thames as described by Wordsworth:

> The river wanders at its own sweet will.

Horace Walpole in his lively and pleasant little work on Modern Gardening almost anticipates this thought. In commending Kent's style of landscape-gardening he observes: "*The gentle stream was taught to serpentize at its pleasure.*"

[036] This Palm-house, "the glory of the gardens," occupies an area of 362 ft. in length; the centre is an hundred ft. in width and 66 ft. in height.

It must charm a Native of the East on a visit to our country, to behold such carefully cultured specimens, in a great glass-case in England, of the trees called by Linnaeus "the Princes of the vegetable kingdom," and which grow so wildly and in such abundance in every corner of Hindustan. In this conservatory also are the banana and plantain. The people of England are in these days acquainted, by touch and sight, with almost all the trees that grow in the several quarters of the world. Our artists can now take sketches of foreign plants without crossing the seas. An allusion to the Palm tree recals some criticisms on Shakespeare's botanical knowledge.

"Look here," says *Rosalind*, "what I found on a palm tree." "A palm tree in the forest of Arden," remarks Steevens, "is as much out of place as a lioness in the subsequent scene." Collier tries to get rid of the difficulty by suggesting that Shakespeare may have written *plane tree*. "Both the remark and the suggestion," observes Miss Baker, "might have been spared if those gentlemen had been aware that in the counties bordering on the Forest of Arden, the name of an exotic tree is transferred to an indigenous one." The *salix caprea*, or goat-willow, is popularly known as the "palm" in Northamptonshire, no doubt from having been used for the decoration of churches on Palm Sunday--its graceful yellow blossoms, appearing at a time when few other trees have put forth a leaf, having won for it that distinction. Clare so calls it:--

> "Ye leaning palms, that seem to look Pleased o'er your image in the brook."

That Shakespeare included the willow in his forest scenery is certain, from another passage in the same play:--

> "West of this place, down in the neighbour bottom. The *rank of osiers* by the murmuring stream, Left on your right hand brings you to the place."

The customs and amusements of Northamptonshire, which are frequently noticed in these volumes, were identical with those of the neighbouring county of Warwick, and, in like manner illustrate very clearly many passages in the great dramatist.--*Miss Baker's "Glossary of Northamptonshire Words." (Quoted by the London Athenaeum.)*

[037] Mrs. Hemans once took up her abode for some weeks with Wordsworth at Rydal Mount, and was so charmed with the country around, that she was induced to take a cottage called *Dove's Nest*, which over-looked the lake of Windermere. But tourists and idlers so haunted her retreat and so worried her for autographs and Album contributions, that she was obliged to make her escape. Her little cottage and garden in the village of Wavertree, near Liverpool, seem to have met the fate which has befallen so many of the residences of the poets. "Mrs. Hemans's little flower-garden" (says a late visitor) "was no more--but rank grass and weeds sprang up luxuriously; many of the windows were broken; the entrance gate was off its hinges: the vine in front of the house trailed along the ground, and a board, with '*This house to let*' upon it, was nailed on the door. I entered the deserted garden and looked into the little parlour--once so full of taste and elegance; it was gloomy and cheerless. The paper was spotted with damp, and spiders had built their webs in the corner. As I mused on the uncertainty of human life, I exclaimed with the eloquent Burke,--'What shadows we are, and what shadows we pursue!'"

The beautiful grounds of the late Professor Wilson at Elleray, we are told by Mr. Howitt in his interesting "*Homes and Haunts of the British Poets*" have also been sadly changed. "Steam," he says, "as little as time, has respected the sanctity of the poet's home, but has drawn its roaring iron steeds opposite to its gate and has menaced to rush through it and lay waste its charmed solitude. In plain words, I saw the stages of a projected railway running in an ominous line across the very lawn and before the windows of Elleray." I believe the whole place has been purchased by a Railway Company.

[038] In Churton's *Rail Book of England*, published about three years ago, Pope's Villa is thus noticed--"Not only was this temple of the Muses--this abode of genius--the resort of the learned and the wittiest of the land--levelled to the earth, but all that the earth produced to remind posterity of its illustrious owner, and identify the dead with the living strains he has bequeathed to us, was plucked up by the roots and scattered to the wind." On the authority of William Hewitt I have stated on an earlier page that some splendid Spanish chesnut trees and some elms and cedars planted by Pope at Twickenham were still in existence. But Churton is a later authority. Howitt's book was published in 1847.

[039]*One would have thought &c.* See the garden of Armida, as described by Tasso, C. xvi. 9, &c.

> "In lieto aspetto il bel giardin s'aperse &c."

Here was all that variety, which constitutes the nature of beauty: hill and dale, lawns and crystal rivers, &c.

> "And, that which all faire works doth most aggrace, "The art, which all that wrought, appearéd in no place."

Which is literally from Tasso, C, xvi 9.

> "E quel, che'l bello, e'l caro accresce à l'opre, "L'arte, che tutto fa, nulla si scopre."

The next stanza is likewise translated from Tasso, C. xvi 10. And, if the reader likes the comparing of the copy with the original, he may see many other beauties borrowed from the Italian poet. The fountain, and the two bathing damsels, are taken from Tasso, C. xv, st. 55, &c. which he calls, *Il fonte del riso*. UPTON.

[040] Cowper was evidently here thinking rather of Milton than of Homer.

> *Flowers of all hue*, and without thorns the rose.

Paradise Lost.

Pope translates the passage thus;

> Beds of all various *herbs*, for ever green, In beauteous order terminate the scene.

Homer referred to pot-herbs, not to flowers of all hues. Cowper is generally more faithful than Pope, but he is less so in this instance. In the above description we have Homer's highest conception of a princely garden:--in five acres were included an orchard, a vineyard, and some beds of pot-herbs. Not a single flower is mentioned, by the original author, though his translator has been pleased to steal some from the garden of Eden and place them on "the verge extreme" of the four acres. Homer of course meant to attach to a Royal residence as Royal a garden; but as Bacon says, "men begin to build stately sooner than to garden finely, as if gardening were the greater perfection." The mansion of Alcinous was of brazen walls with golden columns; and the Greeks and Romans had houses that were models of architecture when their gardens exhibited no traces whatever of the hand of taste.

[041]

> *And over him, art stryving to compayre With nature, did an arber greene dispied*

This whole episode is taken from Tasso, C. 16, where Rinaldo is described in dalliance with Armida. The bower of bliss is her garden

> "Stimi (si misto il culto e col negletto) "Sol naturali e gli ornamenti e i siti, "Di natura arte par, che per diletto "L'imitatrice sua scherzando imiti."

See also Ovid, *Met* iii. 157

> "Cujus in extremo est antrum nemorale necessu, "Arte laboratum nulla, simulaverat artem "Ingenio natura fuo nam pumice vivo, "Et lenibus tophis nativum duxerat arcum "Fons sonat a dextra, tenui perlucidas unda "Margine gramineo patulos incinctus hiatus"

UPTON

If this passage may be compared with Tasso's elegant description of Armida's garden, Milton's *pleasant grove* may vie with both.[141] He is, however, under obligations to the sylvan scene of Spenser before us. Mr. J.C. Walker, to whom the literature of Ireland and of Italy is highly indebted, has mentioned to me his surprise that the writers on modern gardening should have overlooked the beautiful pastoral description in this and the two following stanzas.[142] It is worthy a place, he adds, in the Eden of Milton. Spenser, on this occasion, lost sight of the "trim gardens" of Italy and England, and drew from the treasures of his own rich imagination. TODD.

> *And fast beside these trickled softly downe. A gentle stream, &c.*

Compare the following stanza in the continuation of the *Orlando Innamorato*, by Nilcolo degli Agostinti, Lib. iv, C. 9.

> "Ivi è un mormorio assai soave, e basso, Che ogniun che l'ode lo fa addornientare, L'acqua, ch'io dissi gia per entro un sasso E parea che dicesse nel sonare. Vatti riposa, ormai sei stanco, e lasso, E gli augeletti, che s'udian cantare, Ne la dolce armonia par che ogn'un dica, Deh vien, e dormi ne la piaggia, aprica,"

Spenser's obligations to this poem seem to have escaped the notice of his commentators. J.C. WALKER.

[042] The oak was dedicated to Jupiter, and the poplar to Hercules.

[043] *Sicker*, surely; Chaucer spells it *siker*.

[044] *Yode*, went.

[045] *Tabreret*, a tabourer.

[046] *Tho*, then

[047] *Attone*, at once--with him.

[048] Cato being present on one occasion at the floral games, the people out of respect to him, forbore to call for the usual exposures; when informed of this he withdrew, that the spectators might not be deprived of their usual entertainment.

[049] What is the reason that an easterly wind is every where unwholesome and disagreeable? I am not sufficiently scientific to answer this question. Pope takes care to notice the fitness of the easterly wind for the *Cave of Spleen*.

> No cheerful breeze this sullen region knows, The dreaded east is all the wind that blows.

Rape of the Lock.

[050] One sweet scene of early pleasures in my native land I have commemorated in the following sonnet:--

NETLEY ABBEY.

> Romantic ruin! who could gaze on thee Untouched by tender thoughts, and glimmering dreams Of long-departed years? Lo! nature seems Accordant with thy silent majesty! The far blue hills--the smooth reposing sea-- The lonely forest--the meandering streams-- The farewell summer sun, whose mellowed beams Illume thine ivied halls, and tinge each tree, Whose green arms round thee cling--the balmy air-- The stainless vault above, that cloud or storm 'Tis hard to deem will ever more deform-- The season's countless graces,--all appear To thy calm glory ministrant, and form A scene to peace and meditation dear!

D.L.R.

[051] "I was ever more disposed," says Hume, "to see the favourable than the unfavourable side of things; *a turn of mind which it is more happy to possess, than to be born to an estate of ten thousand a year.*"

[052] So called, because the grounds were laid out in a tasteful style, under the direction of Lord Auckland's sister, the Honorable Miss Eden.

[053] *Songs of the East by Mrs. W.S. Carshore. D'Rozario & Co, Calcutta* 1854.

[054] The lines form a portion of a poem published in *Literary Leaves* in the year 1840.

[055] Perhaps some formal or fashionable wiseacres may pronounce such simple ceremonies *vulgar*. And such is the advance of civilization that even the very chimney-sweepers themselves begin to look upon their old May-day merry-makings as beneath the dignity of their profession. "Suppose now" said Mr. Jonas Hanway to a sooty little urchin, "I were to give you a shilling." "Lord Almighty bless your honor, and thank you." "And what if I were to give you a fine tie-wig to wear on May-day?" "Ah! bless your honor, my master wont let me go out on May-day," "Why not?" "Because, he says, *it's low life.*" And yet the merrie makings on May- day which are now deemed *ungenteel* by chimney-sweepers were once the delight of Princes:--

> Forth goth all the court, both most and least, To fetch the flowres fresh, and branch and blome, And namely hawthorn brought both page and grome, And then rejoicing in their great delite Eke ech at others threw the flowres bright, The primrose, violet, and the gold With fresh garlants party blue and white.

Chaucer.

[056] The May-pole was usually decorated with the flowers of the hawthorn, a plant as emblematical of the spring as the holly is of Christmas. Goldsmith has made its name familiar even to the people of Bengal, for almost every student in the upper classes of the Government Colleges has the following couplet by heart.

> The *hawthorn bush*, with seats beneath the shade, For talking age and whispering lovers made.

The hawthorn was amongst Burns's floral pets. "I have," says he, "some favorite flowers in spring, among which are, the mountain daisy, the harebell, the fox-glove, the wild-briar rose, the budding birch and the hoary hawthorn, that I view and hang over with particular delight."

L.E.L. speaks of the hawthorn hedge on which "the sweet May has showered its white luxuriance," and the Rev. George Croly has a patriotic allusion to this English plant, suggested by a landscape in France.

> 'Tis a rich scene, and yet the richest charm That e'er clothed earth in beauty, lives not here. Winds no green fence around the cultured farm *No blossomed hawthorn shields the cottage dear*: The land is bright; and yet to thine how drear, Unrivalled England! Well the thought may pine For those sweet fields where, each a little sphere, In shaded, sacred fruitfulness doth shine, And the heart higher beats that says; 'This spot is mine.'

[057] On May-day, the Ancient Romans used to go in procession to the grotto of Egeria.

[058] See what is said of palms in a note on page 81.

[059] Phillips's *Flora Historica*.

[060] The word primrose is supposed to be a compound of *prime* and *rose*, and Spenser spells it prime rose

> The pride and prime rose of the rest Made by the maker's self to be admired

The Rev. George Croly characterizes Bengal as a mountainous country--

> There's glory on thy *mountains*, proud Bengal--

and Dr. Johnson in his *Journey of a day*, (Rambler No. 65) charms the traveller in Hindustan with a sight of the primrose and the oak.

"As he passed along, his ears were delighted with the morning song of the bird of paradise; he was fanned by the last flutters of the sinking breeze, and sprinkled with dew by groves of spices, he sometimes contemplated the towering height of the oak, monarch of the hills; and sometimes caught the gentle fragrance of the primrose, eldest daughter of the spring."

In some book of travels, I forget which, the writer states, that he had seen the primrose in Mysore and in the recesses of the Pyrenees. There is a flower sold by the Bengallee gardeners for the primrose, though it bears but small resemblance to the English flower of that name. On turning to Mr. Piddington's Index to the Plants of India I find under the head of *Primula*--Primula denticula--Stuartii--rotundifolia--with the names in the Mawar or Nepaulese dialect.

[061] In strewing their graves the Romans affected the rose; the Greeks amaranthus and myrtle: the funeral pyre consisted of sweet fuel, cypress, fir, larix, yew, and trees perpetually verdant lay silent expressions of their surviving hopes. *Sir Thomas Browne.*

[062] The allusion to the cowslip in Shakespeare's description of Imogene must not be passed over here.--

> On her left breast A mole cinque-spotted, like the crimson drop I' the bottom of the cowslip.

[063] The Guelder rose--This elegant plant is a native of Britain, and when in flower, has at first sight, the appearance of a little maple tree that has been pelted with snow balls, and we almost fear to see them melt away in the warm sunshine--*Glenny*.

[064] In a greenhouse

[065] Some flowers have always been made to a certain degree emblematical of sentiment in England as elsewhere, but it was the Turks who substituted flowers for words to such an extent as to

entitle themselves to be regarded as the inventors of the floral language.

[066] The floral or vegetable language is not always the language of love or compliment. It is sometimes severe and scornful. A gentleman sent a lady a rose as a declaration of his passion and a slip of paper attached, with the inscription--"If not accepted, I am off to the war." The lady forwarded in return a mango (man, go!)

[067] No part of the creation supposed to be insentient, exhibits to an imaginative observer such an aspect of spiritual life and such an apparent sympathy with other living things as flowers, shrubs and trees. A tree of the genus Mimosa, according to Niebuhr, bends its branches downward as if in hospitable salutation when any one approaches near to it. The Arabs, are on this account so fond of the "courteous tree" that the injuring or cutting of it down is strictly prohibited.

[068] It has been observed that the defense is supplied in the following line--*want of sense*--a stupidity that "errs in ignorance and not in cunning."

[069] There is apparently so much doubt and confusion is to the identity of the true Hyacinth, and the proper application of its several names that I shall here give a few extracts from other writers on this subject.

Some authors suppose the Red Martagon Lily to be the poetical Hyacinth of the ancients, but this is evidently a mistaken opinion, as the azure blue color alone would decide and Pliny describes the Hyacinth as having a sword grass and the smell of the grape flower, which agrees with the Hyacinth, but not with the Martagon. Again, Homer mentions it with fragrant flowers of the same season of the Hyacinth. The poets also notice the hyacinth under different colours, and every body knows that the hyacinth flowers with sapphire colored purple, crimson, flesh and white bells, but a blue martagon will be sought for in vain. *Phillips' Flora Historica.*

A doubt hangs over the poetical history of the modern, as well as of the ancient flower, owing to the appellation *Harebell* being, indiscriminately applied both to *Scilla* wild Hyacinth, and also to *Campanula rotundifolia, Blue Bell.* Though the Southern bards have occasionally misapplied the word *Harebell* it will facilitate our understanding which flower is meant if we bear in mind as a general rule that that name is applied differently in various parts of the island, thus the Harebell of Scottish writers is the *Campanula*, and the Bluebell, so celebrated in Scottish song, is the wild Hyacinth or *Scilla* while in England the same names are used conversely, the *Campanula* being the Bluebell and the wild Hyacinth the Harebell. *Eden Warwick.*

The Hyacinth of the ancient fabulists appears to have been the corn- flag, (*Gladiolus communis* of botanists) but the name was applied vaguely and had been early applied to the great larkspur (Delphinium Ajacis) on account of the similar spots on the petals, supposed to represent the Greek exclamation of grief *Ai Ai,* and to the hyacinth of modern times.

Our wild hyacinth, which contributes so much to the beauty of our woodland scenery during the spring, may be regarded as a transition species between scilla and hyacinthus, the form and drooping habit of its flower connecting it with the latter, while the six pieces that form the two outer circles, being separate to the base, give it the technical character of the former. It is still called *Hyacinthus non-scriptus--* but as the true hyacinth equally wants the inscription, the name is singularly inappropriate. The botanical name of the hyacinth is *Hyacinthus orientalis* which applies equally to all the varieties of colour, size and fulness.--*W. Hinks.*

[070] Old Gerard calls it Blew Harebel or English *Jacint*, from the French *Jacinthe*.

[071] Inhabitants of the Island of Chios

[072] Supposed by some to be Delphinium Ajacis or Larkspur. But no one can discover any letters on the Larkspur.

[073] Some *savants* say that it was not the *sunflower* into which the lovelorn lass was transformed, but the *Heliotrope* with its sweet odour of vanilla. Heliotrope signifies *I turn towards the sun*. It could not have been the sun flower, according to some authors because that came from Peru and Peru was not known to Ovid. But it is difficult to settle this grave question. As all flowers turn towards the

sun, we cannot fix on any one that is particularly entitled to notice on that account.

[074] Zephyrus.

[075] "A remarkably intelligent young botanist of our acquaintance asserts it as his firm conviction that many a young lady who would shrink from being kissed under the mistletoe would not have the same objection to that ceremony if performed *under the rose.*"--*Punch*.

[076] Mary Howitt mentions that amongst the private cultivators of roses in the neighbourhood of London, the well-known publisher Mr. Henry S. Bohn is particularly distinguished. In his garden at Twickenham one thousand varieties of the rose are brought to great perfection. He gives a sort of floral fete to his friends in the height of the rose season.

[077] The learned dry the flower of the Forget me not and flatten it down in their herbals, and call it, *Myosotis Scorpioides--Scorpion shaped mouse's ear*! They have been reproached for this by a brother savant, Charles Nodier, who was not a learned man only but a man of wit and sense.--*Alphonse Karr*.

[078] The Abbé Molina in his History of Chili mentions a species of basil which he calls *ocymum salinum*: he says it resembles the common basil, except that the stalk is round and jointed; and that though it grows sixty miles from the sea, yet every morning it is covered with saline globules, which are hard and splendid, appearing at a distance like dew; and that each plant furnishes about an ounce of fine salt every day, which the peasants collect and use as common salt, but esteem it superior in flavour.--*Notes to Darwin's Loves of the Plants*.

[079] The Dutch are a strange people and of the most heterogeneous composition. They have an odd mixture in their nature of the coldest utilitarianism and the most extravagant romance. A curious illustration of this is furnished in their tulipomania, in which there was a struggle between the love of the substantial and the love of the beautiful. One of their authors enumerates the following articles as equivalent in money value to the price of one tulip root--"two lasts of wheat--four lasts of rye--four fat oxen--eight fat swine--

twelve fat sheep--two hogsheads of wine--four tons of butter--one thousand pounds of cheese--a complete bed--a suit of clothes--and a silver drinking cup."

[080]*Maun*, must

[081]*Stoure*, dust

[082]*Weet*, wetness, rain

[083]*Glinted*, peeped

[084]*Wa's*, walls.

[085]*Bield*, shelter

[086]*Histie*, dry

[087]*Stibble field*, a field covered with stubble--the stalks of corn left by the reaper.

[088]*The origin of the Daisy*--When Christ was three years old his mother wished to twine him a birthday wreath. But as no flower was growing out of doors on Christmas eve, not in all the promised land, and as no made up flowers were to be bought, Mary resolved to prepare a flower herself. To this end she took a piece of bright yellow silk which had come down to her from David, and ran into the same, thick threads of white silk, thread by thread, and while thus engaged, she pricked her finger with the needle, and the pure blood stained some of the threads with crimson, whereat the little child was much affected. But when the winter was past and the rains were come and gone, and when spring came to strew the earth with flowers, and the fig tree began to put forth her green figs and the vine her buds, and when the voice or the turtle was heard in the land, then came Christ and took the tender plant with its single stem and egg shaped leaves and the flower with its golden centre and rays of white and red, and planted it in the vale of Nazareth. Then, taking up the cup of gold which had been presented to him by the wise men of the East, he filled it at a neighbouring fountain, and watered the flower and breathed upon it. And the plant grew and became the most perfect of plants, and it flowers in every meadow, when the snow disappears, and is itself the snow of spring, delighting the young heart and enticing the old men from the village to the fields. From then until now this flower has conti-

nued to bloom and although it may be plucked a hundred times, again it blossoms--*Colshorn's Deutsche Mythologie furs Deutsche Volk.*

[089] The Gorse is a low bush with prickly leaves growing like a juniper. The contrast of its very brilliant yellow pea shaped blossoms with the dark green of its leaves is very beautiful. It grows in hedges and on commons and is thought rather a plebeian affair. I think it would make quite an addition to our garden shrubbery. Possibly it might make as much sensation with us (Americans) as our mullein does in foreign green-houses,--*Mrs. Stowe.*

[090] George Town.

[091] The hill trumpeter.

[092] Nutmeg and Clove plantations.

[093] Leigh Hunt, in the dedication of his *Stories in Verse* to the Duke of Devonshire speaks of his Grace as "the adorner of the country with beautiful gardens, and with the far-fetched botany of other climates; one of whom it may be said without exaggeration and even without a metaphor, that his footsteps may be traced in flowers."

[094] The following account of a newly discovered flower may be interesting to my readers. "It is about the size of a walnut, perfectly white, with fine leaves, resembling very much the wax plant. Upon the blooming of the flower, in the cup formed by the leaves, is the exact image of a dove lying on its back with its wings extended. The peak of the bill and the eyes are plainly to be seen and a small leaf before the flower arrives at maturity forms the outspread tail. The leaf can be raised or shut down with the finger without breaking or apparently injuring it until the flower reaches its bloom, when it drops,"--*Panama Star.*

[095] Signifying the *dew of the sea*. The rosemary grows best near the sea-shore, and when the wind is off the land it delights the home-returning voyager with its familiar fragrance.

[096] Perhaps it is not known to *all* my readers that some flowers not only brighten the earth by day with their lovely faces, but emit light at dusk. In a note to Darwin's *Loves of the Plants* it is stated that the daughter of Linnaeus first observed the Nasturtium to throw

out flashes of light in the morning before sunrise, and also during the evening twilight, but not after total darkness came on. The philosophers considered these flashes to be electric. Mr. Haggren, Professor of Natural History, perceived one evening a faint flash of light repeatedly darted from a marigold. The flash was afterwards often seen by him on the same flower two or three times, in quick succession, but more commonly at intervals of some minutes. The light has been observed also on the orange, the lily, the monks hood, the yellow goats beard and the sun flower. This effect has sometimes been so striking that the flowers have looked as if they were illuminated for a holiday.

Lady Blessington has a fanciful allusion to this flower light. "Some flowers," she says, "absorb the rays of the sun so strongly that in the evening they yield slight phosphoric flashes, may we not compare the minds of poets to those flowers which imbibing light emit it again in a different form and aspect?"

[097] The Shan and other Poems

[098] My Hindu friend is not answerable for the following notes.

[099]

> And infants winged, who mirthful throw Shafts rose-tipped from nectareous bow.

Kam Déva, the Cupid of the Hindu Mythology, is thus represented. His bow is of the sugar cane, his string is formed of wild bees, and his arrows are tipped with the rose.--*Tales of the Forest*.

[100] In 1811 this plant was subjected to a regular set of experiments by Dr. G. Playfair, who, with many of his brethren, bears ample testimony of its efficacy in leprosy, lues, tenia, herpes, dropsy, rheumatism, hectic and intermittent fever. The powdered bark is given in doses of 5-6 grains twice a day.--*Dr. Voight's Hortus Suburbanus Calcuttensis*.

[101] It is perhaps of the Flax tribe. Mr. Piddington gives it the Sanscrit name of *Atasi* and the Botanical name *Linum usitatissimum*.

[102] Roxburgh calls it "intensely fragrant."

[103] Sometimes employed by robbers to deprive their victims of the power of resistance. In a strong dose it is poison.

[104] It is said to be used by the Chinese to blacken their eyebrows and their shoes.

[105]*Mirábilis jálapa,* or Marvel of Peru, is called by the country people in England *the four o'clock flower*, from its opening regularly at that time. There is a species of broom in America which is called the American clock, because it exhibits its golden flowers every morning at eleven, is fully open by one and closes again at two.

[106] Marvell died in 1678; Linnaeus died just a hundred years later.

[107] This poem (*The Sugar Cane*) when read in manuscript at Sir Joshua Reynolds's, had made all the assembled wits burst into a laugh, when after much blank-verse pomp the poet began a paragraph thus.--

"Now, Muse, let's sing of rats."

And what increased the ridicule was, that one of the company who slyly overlooked the reader, perceived that the word had been originally *mice* and had been altered to *rats* as more dignified.-- *Boswell's Life of Johnson.*

[108] Hazlitt has a pleasant essay on a garden *Sun-dial*, from which I take the following passage:--

Horas non numero nisi serenas--is the motto of a sun dial near Venice. There is a softness and a harmony in the words and in the thought unparalleled. Of all conceits it is surely the most classical. "I count only the hours that are serene." What a bland and care-dispelling feeling! How the shadows seem to fade on the dial plate as the sky looms, and time presents only a blank unless as its progress is marked by what is joyous, and all that is not happy sinks into oblivion! What a fine lesson is conveyed to the mind--to take no note of time but by its benefits, to watch only for the smiles and neglect the frowns of fate, to compose our lives of bright and gentle moments, turning always to the sunny side of things, and letting the rest slip from our imaginations, unheeded or forgotten! How different from the common art of self tormenting! For myself, as I rode along the Brenta, while the sun shone hot upon its sluggish, slimy waves, my sensations were far from comfortable, but the reading this inscription on the side of a glaring wall in an instant restored me to myself, and still, whenever I think of or repeat it, it has the power of wafting me into the region of pure and blissful abstraction.

[109] These are the initial letters of the Latin names of the plants, they will be found at length on the lower column.

[110] Hampton Court was laid out by Cardinal Wolsey. The labyrinth, one of the best which remains in England, occupies only a quarter of an acre, and contains nearly a mile of winding walks. There is an adjacent stand, on which the gardener places himself, to extricate the adventuring stranger by his directions. Switzer condemns this plan for having only four stops and gives a plan for one with twenty.--*Loudon.*

[111] The lower part of Bengal, not far from Calcutta, is here described

[112] Sir William Jones states that the Brahmins believe that the *blue* champac flowers only in Paradise, it being yellow every where else.

[113] The wild dog of Bengal

[114] The elephant.

[115] Even Jeremy Bentham, the great Utilitarian Philosopher, who pronounced the composition and perusal of poetry a mere amusement of no higher rank than the game of Pushpin, had still something of the common feeling of the poetry of nature in his soul. He says of himself--"*I was passionately fond of flowers from my youth, and the passion has never left me.*" In praise of botany he would sometimes observe, "*We cannot propagate stones*:" meaning that the mineralogist cannot circulate his treasures without injuring himself, but the botanist can multiply his specimens at will and add to the pleasures of others without lessening his own.

[116] A man of a polite imagination is let into a great many pleasures that the vulgar are not capable of receiving. He can converse

with a picture and find an agreeable companion in a statue. He meets with a secret refreshment in a description, *and often feels a greater satisfaction in the prospect of fields and meadows, than another does in the possession.--Spectator.*

[117] Kent died in 1748 in the 64th year of his age. As a painter he had no great merit, but many men of genius amongst his contemporaries had the highest opinion of his skill as a Landscape-gardener. He sometimes, however, carried his love of the purely natural to a fantastic excess, as when in Kensington-garden he planted dead trees to give an air of wild truth to the landscape.

> In Esher's peaceful grove, Where Kent and nature strove for Pelham's love,

this landscape-gardener is said to have exhibited a very remarkable degree of taste and judgment. I cannot resist the temptation to quote here Horace Walpole's eloquent account of Kent: "At that moment appeared Kent, painter and poet enough to taste the charms of landscape, bold and opinionative enough to dare and to dictate, and born with a genius to strike out a great system from the twilight of imperfect essays. He leaped the fence and saw that all nature was a garden[143]. He felt the delicious contrast of hill and valley changing imperceptibly into each other, tasted the beauty of the gentle swell, or concave swoop, and remarked how loose groves crowned an easy eminence with happy ornament, and while they called in the distant view between their graceful stems, removed and extended the perspective by delusive comparison."--*On Modern Gardening*.

[118] When the rage for a wild irregularity in the laying out of gardens was carried to its extreme, the garden paths were so ridiculously tortuous or zig-zag, that, as Brown remarked, a man might put one foot upon *zig* and the other upon *zag*.

[119] The natives are much too fond of having tanks within a few feet of their windows, so that the vapours from the water go directly into the house. These vapours are often seen hanging or rolling over the surface of the tank like thick wreaths of smoke.

[120] Broken brick is called *kunkur*, but I believe the real kunkur is real gravel, and if I am not mistaken a pretty good sort of gravel, formed of particles of red granite, is obtainable from the Rajmahal hills.

[121] Pope in his well known paper in the *Guardian* complains that a citizen is no sooner proprietor of a couple of yews but he entertains thoughts of erecting them into giants, like those of Guildhall. "I know an eminent cook," continues the writer, "who beautified his country seat with a coronation dinner in greens, where you see the Champion flourishing on horseback at one end of the table and the Queen in perpetual youth at the other."

When the desire to subject nature to art had been carried to the ludicrous extravagances so well satirized by Pope, men rushed into an opposite extreme. Uvedale Price in his first rage for nature and horror of art, destroyed a venerable old garden that should have been respected for its antiquity, if for nothing else. He lived to repent his rashness and honestly to record that repentance. Coleridge, observed to John Sterling, that "we have gone too far in destroying the old style of gardens and parks." "The great thing in landscape gardening" he continued "is to discover whether the scenery is such that the country seems to belong to man or man to the country."

[122] In England it costs upon the average about 12 shillings or six rupees to have a tree of 30 feet high transplanted.

[123] I believe the largest leaf in the world is that of the Fan Palm or Talipot tree in Ceylon. "The branch of the tree," observes the author of *Sylvan Sketches*, "is not remarkably large, but it bears a leaf large enough to cover twenty men. It will fold into a fan and is then no bigger than a man's arm."

[124] Southey's Common-Place Book.

[125] The height of a full grown banyan may be from sixty to eighty feet; and many of them, I am fully confident, cover at least two acres.--*Oriental Field Sports*.

There is a banyan tree about five and twenty miles from Berhampore, remarkable for the height of the lower branches from the ground. A man standing up on the houdah of an elephant may pass under it without touching the foliage.

A tree has been described as growing in China of a size so prodigious that one branch of it only will so completely cover two hundred sheep that they cannot be perceived by those who approach the tree, and another so enormous that eighty persons can scarcely embrace the trunk.--*Sylvan Sketches*.

[126] This praise is a little extravagant, but the garden is really very tastefully laid out, and ought to furnish a useful model to such of the people of this city as have spacious grounds. The area of the garden is about two hundred and fifty nine acres. This garden was commenced in 1768 by Colonel Kyd. It then passed to the care of Dr. Roxburgh, who remained in charge of it from 1793 to the date of his death 1813.

[127] Alphonse Karr, bitterly ridicules the Botanical *Savants* with their barbarous nomenclature. He speaks of their mesocarps and quinqueloculars infundibuliform, squammiflora, guttiferas monocotyledous &c. &c. with supreme disgust. Our English poet, Wordsworth, also used to complain that some of our familiar English names of flowers, names so full of delightful associations, were beginning to be exchanged even in common conversation for the coldest and harshest scientific terms.

[128]*The Hand of Eve*--the handiwork of Eve.

[129]*Without thorn the rose*: Dr. Bentley calls this a puerile fancy. But it should be remembered, that it was part of the curse denounced upon the Earth for Adam's transgression, that it should bring forth thorns and thistles. *Gen*. iii. 18. Hence the general opinion has prevailed, that there were *no thorns* before; which is enough to justify a poet, in saying "*the rose was without thorn*."--NEWTON.

[130] See page 188. My Hindu friend is not responsible for the selection of the following notes.

[131] Birdlime is prepared from the tenacious milky juice of the Peepul and the Banyan. The leaves of the Banyan are used by the Bramins to eat off, for which purpose they are joined together by inkles. Birds are very fond of the fruit of the Peepul, and often drop

the seeds in the cracks of buildings, where they vegetate, occasioning great damage if not removed in time.--*Voight*.

[132] The ancient Greeks and Romans also married trees together in a similar manner.--*R*.

[133] The root of this plant, (*Euphorbia ligularia,*) mixed up with black pepper, is used by the Natives against snake bites.--*Roxburgh*.

[134] Coccos nucifera, the *root* is sometimes masticated instead of the Betle-nut. In Brazil, baskets are made of the *small fibres*. The *hard case of the stem* is converted into drums, and used in the construction of huts. The lower part is so hard as to take a beautiful polish, when it resembles agate. The reticulated substance at base of the leaf is formed into cradles, and, as some say, into a coarse kind of cloth. The *unexpanded terminal bud* is a delicate article of food. The *leaves* furnish thatch for dwellings, and materials for fences, buckets, and baskets; they are used for writing on, and make excellent torches; potash in abundance is yielded by their ashes. The *midrib of the* leaf serves for oars. The *juice of the flower and stems* is replete with sugar, and is fermented into excellent wine, or distilled into arrack, or the sugary part is separated as Jagary. The tree is cultivated in many parts of the Indian islands, for the sake not only of the sap and *milk* it yields, but for the *kernel* of its fruit, used both as food and for culinary purposes, and as affording a large proportion of *oil* which is burned in lamps throughout India, and forms also a large article of export to Europe. The fibrous and uneatable rind of the fruit is not only used to polish furniture and to scour the floors of rooms, but is manufactured into a kind of cordage, (*Koir*) which is nearly equal in strength to hemp, and which Roxburgh designates as the very best of all materials for cables, on account of its great elasticity and strength. The sap of this as well as of other palms is found to be the simplest and easiest remedy that can be employed for removing constipation in persons of delicate habit, especially European females.--*Voigt's Suburbanus Calcuttensis*.

[135] The root is bitter, nauseous, and used in North America as anthelmintic. *A. Richard*.

[136] Of one species of tulsi (*Babooi-tulsi*) the seeds, if steeped in water, swell into a pleasant jelly, which is used by the Natives in

cases of catarrh, dysentry, chronic diarrhoea &c. and is very nourishing and demulcent--*Voigt*.

[137] This list is framed from such as were actually grown by the author between 1837 and the present year, from seed received chiefly through the kindness of Captain Kirke.

[138] The native market gardens sell Madras roses at the rate of thirteen young plants for the rupee. Mrs. Gore tells us that in London the most esteemed kinds of old roses are usually sold by nurserymen at fifty shillings a hundred the first French and other varieties seldom exceed half a guinea a piece.

[139] I may add to Mr. Speede's list of Roses the *Banksian Rose*. The flowers are yellow, in clusters, and scentless. Mrs. Gore says it was imported into England from the Calcutta Botanical Garden, it is called *Wong moue heong*. There is another rose also called the *Banksian Rose* extremely small, very double, white, expanding from March till May, highly scented with violets. The *Rosa Brownii* was brought from Nepaul by Dr. Wallich. A very sweet rose has been brought into Bengal from England. It is called *Rosa Peeliana* after the original importer Sir Lawrence Peel. It is a hybrid. I believe it is a tea scented rose and is probably a cross between one of that sort and a common China rose, but this is mere conjecture. The varieties of the tea rose are now cultivated by Indian malees with great success. They sell at the price of from eight annas to a rupee each. A variety of the Bengal yellow rose, is now comparatively common. It fetches from one to three rupees, each root. It is known to the native gardeners by the English name of "*Yellow Rose*". Amongst the flowers introduced here since Mr. Speede's book appeared, is the beautiful blue heliotrope which the natives call *kala heliotrope*.

[140]

> He gains all points who pleasingly confounds, Surprizes, varies, and conceals the bounds.

[141] The following is the passage alluded to by Todd

> A pleasant grove With chant of tuneful birds resounding loud, Thither he bent his way, determined there To rest at

noon, and entered soon the shade, High roofed, and walks beneath and alleys brown, That opened in the midst a woody scene, Nature's own work it seemed (nature taught art) And to a superstitious eye the haunt Of wood gods and wood nymphs.

Paradise Regained, Book II

[142] The following stanzas are almost as direct translations from Tasso as the two last stanzas in the words of Fairfax on page 111:--

> The whiles some one did chaunt this lovely lay;-- Ah! see, whoso fayre thing doest faine to see, In springing flowre the image of thy day! Ah! see the virgin rose, how sweetly shee Doth first peepe forth with bashful modesty; That fairer seems the less you see her may! Lo! see soone after how more bold and free Her baréd bosome she doth broad display; Lo! see soone after how she fades and falls away! So passeth, in the passing of a day, Of mortal life, the leaf, the bud, the flowre, Ne more doth florish after first decay, That erst was sought, to deck both bed and bowre Of many a lady and many a paramoure! Gather therefore the rose whilest yet is prime For soone comes age that will her pride deflowre; Gather the rose of love, whilest yet is time Whilest loving thou mayst loved be with equal crime[144]

Fairie Queene, Book II. Canto XII.

[143] I suppose in the remark that Kent leapt the fence, Horace Walpole alludes to that artist's practice of throwing down walls and other boundaries and sinking fosses called by the common people *Ha! Ha's!/* to express their astonishment when the edge of the fosse brought them to an unexpected stop.

Horace Walpole's History of Modern Gardening is now so little read that authors think they may steal from it with safety. In the *Encyclopaedia Britannica* the article on Gardening is taken almost verbatim from it, with one or two deceptive allusions such as--"*As Mr. Walpole observes*"--"*Says Mr. Walpole,*" &c. but there is nothing to mark where Walpole's observations and sayings end, and the

Encyclopaedia thus gets the credit of many pages of his eloquence and sagacity. The whole of Walpole's *History of Modern Gardening* is given piece-meal as an original contribution to *Harrrison's Floricultural Cabinet*, each portion being signed CLERICUS.

[144] Perhaps Robert Herrick had these stanzas in his mind's ear when he wrote his song of

> Gather ye rosebuds while ye may Old time is still a flying; And this same flower that smiles to-day To-morrow will be dying.
>
> Then be not coy, but use your time; And while ye may, so marry: For having lost but once your prime You may for ever tarry.